RECENT TECHNIQUES IN POULTRY FARMING AND FEED FORMULATIONS

RECENT TECHNIQUES IN POULTRY FARMING AND FEED FORMULATIONS

Ganesh Yadav

RANDOM PUBLICATIONS

NEW DELHI - 110 002 (INDIA)

Recent Techniques in Poultry Farming and Feed Formulations

ISBN 978-93-51116-99-8

Published in 2015 in India by

RANDOM PUBLICATIONS

4376-A/4B, Gali Murari Lal, Ansari Road

New Delhi-110 002

Phone: +9111-43580356, 23289044

E-mail: randomexports@gmail.com; sales@randompublications.com; info@randompublications.com

Reprinted 2018

Type Setting by: Friends Media, Delhi-110089

Digitally Printed at : Replika Press Pvt. Ltd.

Preface

As farming became more specialized, many farms kept flocks too large to be fed in this way, and nutritionally complete poultry feed was developed. Modern feeds for poultry consists largely of grain, protein supplements such as soybean oil meal, mineral supplements, and vitamin supplements. The quantity of feed, and the nutritional requirements of the feed, depend on the weight and age of the poultry, their rate of growth, their rate of egg production, the weather, and the amount of nutrition the poultry obtain from foraging. This results in a wide variety of feed formulations. The substitution of less expensive local ingredients introduces additional variations.

Feed formulation is the process of quantifying the amounts of feed ingredients that need to be combined to form a single uniform mixture (diet) for poultry that supplies all of their nutrient requirements. Since feed accounts for 65-75% of total live production costs for most types of poultry throughout the world, a simple mistake in diet formulation can be extremely expensive for a poultry producer. Most large-scale poultry farmers have their own nutritionists and feed mills, whereas small operations usually depend on consultant nutritionists and commercial feed mills for their feeds. It is therefore essential that formulations are accurate because once feeds are formulated and manufactured, it is often too late to remedy any mistakes or inaccuracies without incurring significant expenses. Feed formulation is both a science and an art, requiring knowledge of feed and poultry, and some patience and innovation when using formulae. Typical formulations indicate the amounts of each ingredient that should be included in the diet, and then provide the concentration of nutrients (composition) in the diet. The nutrient composition of the diet will indicate the adequacy of the diet for the particular class of poultry for which it is prepared. It is common to show the energy value in metabolisable energy (kcal or MJ ME/kg feed) and protein content of the diet but comprehensive information on concentrations of mineral elements and digestible amino acids are also provided. Digestible amino acids often include not just the first limiting amino acid, methionine, but also most of the ten essential amino acids. A number of databases are available to provide information on the digestible amino acid contents of various poultry feed ingredients. Feed formulation, often referred to as least cost formulation, is the process of matching the nutrient requirements of a class of animals with the nutrient contents of the available ingredients (raw materials) in an economic manner. Although some feed mills produce test diets for evaluation in the laboratory or in feeding trials to confirm the adequacy of the diet, the most important preparation for accurate and economic formulation is to test the chemical composition of the ingredients available for use. Most feed mills today have their own quality control (QC) laboratories. The future development of the poultry industry in many regions of the world depends to a large extent on the availability of feedstuffs in those areas that are suitable or can be made suitable for use in *poultry feeds*.

This book provides detailed scientific information on the processing of agro-waste material to provide inexpensive alternatives to non-traditional feedstuffs for use in poultry and farm animal nutrition.

I thank all members of my team who have helped in the preparation of the book. My special thanks go to "Random Publications" who have published the book.

— *Ganesh Yadav*

Contents

Chapter 1

Introduction

Local chickens are kept in many parts of the world irrespectively of the climate, traditions, life standard, or religious taboos relating to consumption of eggs and chicken meat like those for pig meat (Tadelle, 2003). To the poor majority in rural areas, local chickens serve as an immediate source of meat and income when money is needed for urgent family needs (Ekue *et al.*, 2002). It constitutes a significant contribution to human livelihood and contributes significantly to food security (Gondwe, 2004). Women and sometimes youths are the mostly involved in keeping these chickens. The local chickens are known for various merits.

Most important, they are known for their adaptation superiority in terms of their resistance to endemic diseases and other harsh environmental conditions. However, local chickens are poor performers in terms of growth rate (hence meat production) and egg production. Most of them are of small adult size and lay small sized eggs when compared to improved commercial broiler or layer birds respectively (Pedersen, 2002; Gondwe, 2004). What is generally referred to as local chickens is a pool of heterogeneous individuals which differ in adult body size, weight and plumage.

They are of several ecotypes that are distinct. Their performance vary considerably and no single ecotype meets the attributes of good egg traits, fertility, hatchability, survivability, high growth rate, heavy weight at slaughter and high egg production. Fortunately, their genetic diversity could be exploited to improve their productivity. It is therefore a laudable proposition that more attention should be given to the genetic improvement and development of the local chicken in order to ameliorate the present acute animal protein shortage to many poor societies around the globe. One way of improving the local chicken is by cross breeding with improved commercial breeds. In Nigeria, cross breeding local fowls to commercial Rhode Island Red (RIR) chicken produced Fulani-ecotype chicken that is superior to other local ecotypes within Nigeria in terms of egg traits, hatchability, growth performance and live weight (Atteh, 1999; Fayeye 2005). Such improvements provide potentially good ecotypes for meat and egg production and could thus help to develop improved local strains. Studies on improvement of local chickens are rarely reported in other parts of the world including Tanzania. Stemming on the importance of local chicken to the economy of the poor majority in Tanzania, this study was designed to gather preliminary information on the feasibility of improving the local chicken by cross breeding with the commercial RIR. The study explored and compared egg traits, fertility, hatchability, chick hatch weight, and chick survivability for local, RIR, and crossbred chickens.

Materials and Methods

Source of Chickens and Sample Size

A total of 250 (200 females and 50 males) day old chicks of exotic layer breed, Rhode Island Red (RIR), purchased from a local agent were raised to serve a breeding purpose. To obtain local chickens and to study the egg traits of these birds, a total of 6752 local chicken eggs collected from Morogoro urban (2803), villages around Mgeta (3024), and Mamvisi village of Kidete ward, Kilosa (925) (all in Morogoro region, Tanzania) were sorted, incubated, hatched and the chicks grown as local chicken breeding stock. Crossbred chickens were obtained by crossing RIR layer cocks to local hens and vice versa.

Egg Measurements

All eggs for incubation were sorted against cracks, morphological deformities, and dirty (soiled) before acquiring egg weight, egg length, egg breadth, and egg volume. A total of 1382, 1523, and 1476 local, RIR, and crossbred chicken eggs respectively were assessed. Eggs were weighed to the nearest 0.10 gram on a digital scale while egg length and egg breadth were measured to the nearest 0.10 cm using a pair of vernier callipers (GT Tools, Japan). The values of the egg length (L) and egg breadth (B) were used to determine the egg volume (V) (cm3) using Hoyt's (1979) equation (V=Kv*LB2) where the estimated volume coefficient (Kv=0.507) is applicable to all eggs which are not very pointed.

Incubation

Eggs of medium size were incubated in an automated electrical incubator of 1350 egg capacity (Kalambo ET, Kibaha, Tanzania) at 37?C and 60-65% relative humidity and turning hourly. Candling was done on day 5 and 18 to determine infertile eggs ('clears') and dead embryos, respectively. The latter were confirmed by breaking the eggs after 21 days. Eggs with living embryos were then transferred to the hatching chamber of the incubator. Hatched chicks and those assisted to hatch by breaking the egg shell were collected, counted and weighed to the nearest 0.10 gram on a digital scale to determine the chick hatch weight.

Fertility and Hatchability

Both fertility and hatchability were determined in 3675 local chicken eggs and 3350 for RIR and crossbred each. Fertility was determined as 100[number of fertile eggs]/number of total eggs set; while hatchability was determined from the formula: Hatchability = 100[number of chicks hatched]/number of fertile eggs set.

Housing, Diet and Disease Control

Hatched chicks were raised on electrically heated brooder for three weeks. The birds were kept intensively and adults were stocked at 10 birds/m2.

Table: *Composition of chicken feeds*

Ingredient	***Percent in diet of chicken feeds***		
	Chick	***Growers***	***Layers***
Maize	38.0	41.0	45.0
Maize bran	10.0	12.0	11.5
Rice bran	17.0	20.0	19.0
Sunflower seed cake	12.5	10.0	10.0
Cotton seed cake	7.0	7.0	5.0
Fish meal	11	5.5	2.0
Layers premix	0.5	0.5	0.5
Limestone	1.0	1.0	2.5
Salt	0.5	0.5	0.5
Bone meal	1.0	1.0	2.5
Methionine	0.5	0.5	0.5
Lysine	1.0	1.0	1.0

Poultry Growing

Avimax transit broiler cage, the multi-tier battery poultry housing system - for hygienic, efficient and successful poultry growing. AviMax transit combines the AviMax broiler cage with an innovative manure belt system that doubles as an efficient new means of broiler transport. The patent pending pivoting floor optimises housing conditions during the growing phase and permits easy moving-out for ready-to-slaughter broilers.

AviMax sliding - the multi-tier broiler battery for hygienic and efficient poultry growing, and for easy and flexible moving-out of the broilers. Fluxx 330 & 360 feed pans for successful broiler feeding - for rearing and growing; satisfies the needs of day-old chicks as well as those of heavy broilers in the final growing phase. Different variants of Fluxx are available to provide an optimal feed pan for every broiler producer worldwide.

Poultry growing with Augermatic poultry feeding system - with Big Pan 330 or Big Pan Plus — our effective system for modern broiler growing. Big Pan 330 is the proven feed pan for the so called ad libitum feeding and Big Pan Plus is ideally suited for controlled feeding of broilers.

Gladiator - the rugged feed pan for successful rearing and growing of turkeys meets the requirements of day-old poults as well as those of heavy turkeys in the final growing period. Feed transport and filling of all pans of a line is done by means of the reliable Auger Matic conveying system. Modern housing equipment from Big Dutchman for successful turkey production.

Breeder management with EMPA 4 - the self-feeder for broiler breeder males and finishing poultry is particularly suitable for supplying feed to males in broiler breeder management. It is also possible to feed pullet layers and turkeys from this pan.

Drinking systems for poultry production - here you find a survey on our nipple drinkers and round drinkers for rearing birds and heavier poultry, so that they always have fresh and clean drinking water. The respective accessories– e.g. cable winches or water connection units – are also part of the complete program.

Poultry growing with amacs broiler - the Agro Management and Control System for modern broiler production and broiler breeder management. For real-time monitoring and control of poultry production units with central data storage incl. analyses per bird, comparison with reference data stored in the system. Visualisation of the present situation inside the poultry house on the farm controller. Optimal poultry climate control with Viper Touch - the flexible climate and production computer for all types of poultry production.

The computer allows for an optimal poultry climate control and is not only characterised by its simple operation via touchscreen, it is also unsurpassed in terms of comfort, speed and performance. In combination with the PC program BigFarmNet-Manager all climate and poultry production data can easily be analysed.

Efficient poultry management with Viper - the modular climate and production computer for an optimal climate and more efficiency in your poultry house. The easy-to-operate Viper registers all important data concerning production, growth, feed and water consumption, mortality and climate for better production performance and a higher yield.

Precision weighing for all types of poultry production - with Big Dutchman poultry scales you are always up-to-date on the weight of your stock! We offer the ideal weighing system, mobile or stationary, for every type of poultry production, be it meat or egg production or breeder management.

Silos and augers for livestock production - feed storage and transport systems. Choose from a wide range of feed bins made of galvanised sheet steel, glass-fibre reinforced plastic or Trevira fabric. Feed can be transported by augers or spirals without any disadvantages, but depends on the conveying capacity and the pitches.

Milling and mixing systems - the comprehensive Big Dutchman product range includes milling and mixing systems. The trend of using home-grown cereals for the production of mixed feed has increased. This allows the production of individual recipes of known quality and also reduces transport costs.

Residue treatment with OptiSec for poultry production - OptiSec is a new manure belt-drying tunnel developed by Big Dutchman, which provides optimum drying (up to 90 % DM) of fresh or pre-dried manure from poultry cages and enriched cages as well as biomass as e.g. fermentation substrates from biogas plants. OptiSec is characterised by its large capacity which is achieved thanks to its large width.

Poultry Housing

Poultry production systems should provide fresh air, clean feed and water, protection against predators, shelter from cold, rain, wind, sun and excessive heat; as well as a source of heat when birds are young. Basically, the birds need to be able to grow, sleep and lay eggs in comfort. The birds should also be free from stress and disease. The basic requirements for poultry housing are:

- Protection from weather
- Protection from predators
- Enough space
- Adequate ventilation
- A clean environment
- Access to dust bathing facilities.

Protection

A good poultry house protects the birds from the elements (weather), predators, injury and theft. The housing must also provide a stable environment in which the birds feel "comfortable" during the day and at night, are protected against potential predators, and are provided with secure nesting boxes.

Chickens require a dry, and largely draft-free house. In cooler climates, this can be accomplished by building a house with windows and/or doors which can be opened for ventilation when necessary. In hotter climates, windows, doors and even walls may not be necessary.

Build the chicken house on higher, well-drained areas in order to prevent prolonged dampness and water saturation of the floor and outside areas.

If necessary construct drainage channels and ditches to allow heavy rain to drain away quickly or prevent water flooding into the area. Allowing an adequate level of space per bird also helps keep the humidity level inside an enclosed chicken house to a minimum.

Complete Confinement - A Poultry House

Keeping chickens totally confined with a fenced and covered run are the best methods of protection against predators. For new constructions, and especially for larger installations, consider laying a concrete floor for the house. This helps to prevent rodents and snakes from entering, and other potential predators from digging under the walls and floors. Windows and doors must be securely covered with wire mesh. However, under these circumstances, chickens (and any other poultry kept under complete confinement) will need to be supplied with ALL of their daily requirements, and the house will need to be kept clean.

Poultry House with Outside Run

With outside runs, bury the wire along the edges of the pen at least 12 inches or 30 cm deep. Lay the buried fence outwards at a slight angle. This stops most predators from digging under the fence. Many predators will tend to dig at the base of a fence and by setting the fence at an angle under the ground. The animal will encounter the fencing and be deterred. If the outside runs are not predator-proof, the poultry will ned to be securely shut in before dark – and opened up again early in the morning, just after sunrise.

Flying predators can be a particular problem for young chicks. To prevent problems with hawks and owls, cover your outside runs with mesh wire or netting. In the absence of netting, a good substitute can be constructed from a mesh or grid made of string, and this will provide excellent protection against flying predators.

Build chicken houses to prevent possible injury to your birds. Remove any loose or projecting wire, nails, or other sharp-edged objects from the house and run. Except for purpose-made perches, remove any other projections where the birds could attempt to perch – they may damage themselves attempting to perch on inappropriate objects.

Chickens need a shelter with both fresh air and sufficient light. In hot climates, a galvanised or corrugated iron roof can make the house very hot during the day. Chickens are unable to sweat. By opening their beaks and panting they remove body heat. This panting can also be a sign of heat stress. Chickens in an excessively hot house suffer from heat stress. They eat less and thus become less productive (fewer eggs and less growth), and at high temperatures may die.

Space Requirements, or Density of Birds Per Unit Area

This is the most important basic principle in housing, as the space available determines the number and type of poultry that can be kept.

Birds need adequate space for movement and exercise as well as areas to nest and roost. Space requirements vary with the species, type or breed of birds that are raised, as well as the type of production system used.

Minimum space requirements are given by a number of sources, and these should be seen as the minimum space requirements – where the birds are supplied with all dietary needs, and do not have to search and forage for feed and water. Two systems of measurement are used in the literature, either number of birds per square metre, or the square feet required per bird. Examples from the literature are given below:

***Table:** Minimum Space Requirements for different poultry*

Type of Bird	*Sq ft/bird inside*	*Sq ft/bird outside runs*
Bantam Chickens	1	4
Laying Hens	1.5	8
Large Chickens	2	10
Quail	1	4
Pheasant	5	25
Ducks	3	15
Geese	6	18

Source: Clauer, P.J. Small Scale Poultry Housing. Small Flock Factsheet, Number 10. Virginia Cooperative Extension. Virginia Polytechnic Institute and State University.

***Table :** Minimum Requirement of Chickens for floor and perch space*

Chicken types	*Floor Space (birds/m²)*	*Floor Space (ft²/ bird)*	*Perch Space (per bird)*
Layer	3	3.6	25 cm (10 in)
Dual Purpose	4	2.7	20 cm (8 in)
Meat	4-5	2.1-2.7	15-20 cm (6-8 in)

Source: FAO (2004). Small-scale Poultry Production: Technical Guide. FAO Animal Production and Health Manual 1. FAO, Rome.

Hen groups are comfortable at a stocking density of up three to four birds per square metre. However, if more space is allowed, a greater variety of behaviour can be expressed. Less space creates stressed social behaviour, allowing disease vulnerability and even cannibalism, with the weaker birds being deprived of feeding or perching space - or more likely both. Individual birds need more room for normal behaviour and adequate exercise than the very high densities currently used in commercial intensive production of both egg laying birds and broiler chickens. Over recent decades, animal welfare concerns have encouraged research on laying cage structures to make designs better suited to the needs of hens, while retaining cost-effectiveness for production.

Ventilation

A good air circulation and movement, without a draft, is essential in any poultry house. Fresh air brings in oxygen while excess moisture, ammonia and carbon dioxide are removed. High levels of carbon dioxide and ammonia may cause significant problems in intensive chicken production, whilst excessive levels of poisonous carbon monoxide may also cause problems. For small poultry houses, windows or vents on one side of the house usually provide plenty of ventilation. In warmer climates a building with open sides is ideal.

Temperature

Heat stress is a significant constraint to successful production and can lead to death. Although chickens can survive temperatures several degrees below freezing, they do not tolerate temperatures over 40°C. This also depends on the relative humidity. Poultry do not possess sweat glands and cool themselves by evaporative cooling via their breath - beaks open and a rapid movement of air. When the humidity is too high, this cooling mechanism does not work well.

Lethal temperatures for most chickens are 46°C upwards, and severe stress sets in above 40°C. In temperate regions, a chicken house can be constructed facing the rising morning sun to gain heat. In the tropics however, an east-west orientation of the length of the building helps to minimise exposure to direct sunlight. Building materials such as tin, corrugated iron or other metal should be avoided for this reason.

Ground cover can also reduce reflected heat. Shade should be provided, especially if there is little air movement or if humidity is high. With no shade, or when confined in higher temperatures, poultry become heat stressed and irritable, and are likely to start pecking at each other. Particularly in younger birds blood is easily drawn, and this can lead to cannibalism.

The effects of heat stress include:

- a progressive reduction in feed intake as ambient temperature rises
- an increase in water consumption in an attempt to lower temperature
- a progressive reduction in growth rate

- disturbances in reproduction (lower egg weight, smaller chicks, reduced sperm concentration and an increased level of abnormal sperm in cocks).

A good supply of clean water must always be supplied. This should be changed each day, and if necessary the supply should be topped up at intervals during the day.

Access to Feed and Water

Feeders and water should be placed conveniently within the outside pen. It is a good idea that these are covered to protect the feed from the rain. Where rodents become a problem, feeders should be taken out at night and stored or hung up in a secure location. Clean fresh water should be provided daily, and if necessary the supply should be topped up at intervals during the day.

Place the bottom of water containers and the top lip of the feeders at the approximately the height of the back of the birds. This will help to keep the feed and water clean and prevent spillage and wastage. When possible, place the water in the outside runs, especially for waterfowl. This helps to keep the humidity level lower inside the poultry house.

Dust Bathing

Dust bathing refers to the pattern of behaviour where chickens scratch in an area of fine, loose, dry soil and then takes a "bathe" in it - allowing the fine soil particles to trickle through the feathers. This is an instinctive action and helps to get rid of external parasites in the feathers and on the skin. Most freely ranging birds will find a dry, sunny spot where they can make their own dustbaths and use these repeatedly. However, if they are in a run where this is not possible, dust-bathing facilities must be provided.

A dust bath can be constructed using a large shallow box filled with fine sand or earth. The addition of ash from the fire provides extra fine material that is ideal for dust bathing - but make sure it is completely cold first!! This dust bath should be placed under cover or provided with its own roof, since it is more or less useless when wet.

Materials and Designs for Poultry Houses

In regions where it rains heavily, the floor should be raised with a generous roof overhang, particularly over the entrance. The raised floor can be made from either solid platform of earth or be constructed as a raised bamboo or similar platform. A raised bamboo platform has the added advantage of providing ventilation under the poultry, which helps cool them in hot weather and also keeps them out of flood water during heavy rain.

The walls of the building can be made of mud or bamboo, and the windows and door of bamboo slats. The house can be freestanding, or attached to other buildings. Such houses are suitable for semi-intensive production systems.

Free Range or Free to Range

There is a big difference between the term "*Free Range*", and the concept of a bird that is *free to range* within its environment. The term "Free Range" although widely used and generally accepted is actually very misleading, and in most European countries for example, has a specific set of recommended minimum criteria. The birds are not actually free to range. To complicate this even further there is another category called "*Traditional Free Range*".

A study by Dawkins et al. (2003) found that chickens prefer ranging areas with trees, they avoid bright sun and that, within their paddocks, they either stay close to the house or they seek tree cover. This is hardly surprising in view of the fact that domestic birds are descended from red jungle fowl, which inhabit the dry forests of southeast Asia and are also known as "bamboo fowl".

A wide open field without any cover is not a preferred habitat. This has important implications for the design of free-range poultry systems and makes it clear that tree cover is something that should be provided to encourage ranging. A fruit orchard, for example, would be an ideal ranging area. The forests where jungle fowl occur frequently consist of thick clumps of bamboo separated by small clearings, so that the birds can see the approach of predators but are rarely far from cover into which they disappear at the sign of danger.

Free Range

A label of "Free Range" on an egg or poultry meat product is a specific marketing term indicating that the product has been produced in compliance with the criteria set-out in the respective marketing regulations. In the UK, and elsewhere in the EU for example, a product described a as Free Range egg must comply with basic management criteria providing the birds with a maximum stocking density inside the poultry house (9 per m^2), as well as a stated feeder space per bird (10 cm of feeder/bird) and drinker space per bird (one drinker/10 birds). In addition birds must have continuous daytime access to open runs which are mainly covered with vegetation and with a maximum stocking density of 2,500 birds per hectare. There are also specific requirements in terms of the age at slaughter.

Optionally, producers may make further provision for the birds, which will in turn lead to a tastier product, grown in a less intensive, more welfare-friendly way. *Traditional Free Range* requirements differ from Free Range by requiring more extensive open-air access, a lower stocking density, and a greater minimum age at slaughter. *Free Range - Total Freedom* has similar requirements, but birds must have unrestricted daytime open-air access.

Traditional Free Range

Traditional free-range typically involves smaller flocks than Free Range and these are often kept in moveable houses. Their position is changed on a regular basis, often daily, so that the birds have access to a more or less constant supply of fresh green vegetation, with some insects, worms and other natural food.

This is similar to systems in North America called Range Poultry, or Pasture Poultry.

From a production viewpoint, the reasons for providing pasture to the birds are to:

- obtain feed from the pasture (young growing vegetation, insects, worms, etc.)
- to improve land fertility (e.g. for subsequent grasing by ruminants)
- to improve bird health
- to improve the general welfare of the birds
- and to provide products for specific markets (normally fetching a higher price).

Poultry obtain nutrients from young, vegetative forage plants. This results, for example, in eggs with a deep yellow colour. Poultry on pasture also forage for seeds and live protein such as worms and insects. Larger chickens will even catch and eat small rodents and lizards if given the chance.

Extensive range poultry production requires a more land and is usually part of a diversified operation with ruminant and this mixed husbandry can be very important in range poultry production. Soil fertility is a major motivation and producers are able to take advantage of the poultry manure to improve their pastures for ruminants.

Free to Range

Poultry that are actually free to range, tend to be the village poultry, or scavenging poultry, kept by smallholders and households in tropical areas. Smallholders in Europe and North America may also keep poultry that are in effect free to range over considerable areas. It has been estimated that 80% of the poultry population in Africa is found in these traditional production systems, sometimes called *low input/output systems*.

Little attention is given to this means of production by authorities and many development programmes even though between 30% to 100% of the animal protein consumed is from this source. These *low input/output* systems have been a traditional component of small farms all over the developing world for centuries and is likely to continue as such in the future.

The flocks are small in size but are an important asset providing their owners with meat and eggs that can be consumed by the family. They are also bartered or sold to provide additional income, and may also be used to fulfil social obligations. Rural poultry provide manure, whilst they are active in pest control. In many countries the birds are owned and managed by women and children.

Improved Management of Free to Range or Scavenging Poultry

There are a number of basic options for improvement of the management of scavenging poultry, with the aim of increasing production. In terms of housing, the birds can either be confined or can remain free-ranging. Improved management is a combination of better feeding and better housing, whilst also paying attention to the health care needs of the birds (especially vaccination against Newcastle Disease).

Birds that are not Confined

The unrestricted free-ranging of poultry is often a problem. They trespass onto neighbouring fields and gardens, and are constantly at risk from predators. Confinement may not be practical because of the additional costs of feed and fencing. Surveillance is only feasible where the very old or very young of the household have time to help. Fencing of vegetable plots is in many cases the best option. Placing more cocks in the village may reduce the movements of the chickens, as cocks and hens of each flock are more likely to keep to their own territory.

However, poultry can be a significant benefit within the vegetable plot if they are given access at the right times during the growing season. Chickens peck at the ground, eating weeds and they can consume large numbers of insect pests.

The birds get plenty of sunlight, exercise and fresh air. The manure helps to fertilize the soil, replacing nutrients so that the next crop will grow strong and healthy.

In planted orchards, or under fruit trees, birds will also clean up windfalls whilst also fertilizing the trees.

For more on the use of chickens within the vegetable plot, including practical methods, see a Radio Script titled "*Chickens fertilize and weed the garden*". A related radio script titled "*Chickens eat ticks on cattle*" is also relevant.

Under a free-to-range or scavenging system, the difference between the amounts of food gathered through scavenging and the total food requirement for maximum production should be balanced with nutrients supplied by a supplementary feed.

To make up a properly balanced supplement, it is of course necessary to know how much and what types of nutrients the birds are obtaining from the wild (technically termed the scavenger feed resource base).

This is not normally known, and it is recommended that the birds be given access (using a free-choice cafeteria system) to three containers (or three compartments of a bamboo stem feeder) of ingredients comprising a protein concentrate, a carbohydrate source (for energy) and a mineral source (mainly for calcium carbonate for egg shell formation for the hen). Poultry should have free access to this cafeteria system for two to three hours in the evening to supplement the day's scavenging.

Backyard Extensive Systems

Under these systems, poultry are housed at night but allowed free-range during the day. They are usually fed grain in the morning and evening to supplement scavenging. Regular provision of food at the house will ensure that the birds return at night. Improvements in housing will provide benefits. If birds feel more comfortable in the nest boxes provided, they are less likely to nest and lay eggs in the bush.

Night-time Roosts for Scavenging or Backyard Poultry

With largely scavenging poultry, a workable system is to confine them in a secure shelter at night, to protect them from natural predators, while allowing them to roam freely around the home compound during the day, when the predators are less active. In this way, the birds will have ready access to their main diet of insects and seeds that they obtain for themselves and the amount of supplementary feed to be provided by the farm family will be reduced.

Any feed of grain or household scraps that is offered should be given inside the shelter. If this is regularly provided in the evening, it will help to train the birds to willingly enter the enclosure before nightfall.

Figure: *Green leafy vegetation, such as outer leaves from cabbage, can be tied in bunches to the edge of a chicken run.*

If the shelter is left open during the day, the birds will be able to seek shelter from excessively hot, cold or wet weather, as needed.

Chickens need green leafy vegetation, and if confined they should be provided with this - preferably on a daily basis. This needs to be secured, and not simply left loose on the floor of the chicken run. Under free-ranging conditions the birds would peck at leaves and pull off pieces, and if leaves are not secured this becomes very difficult.

Increased Confinement: Semi-Intensive Systems

These are a combination of the extensive and intensive systems, where birds are confined to an area with access to shelter. Where confinement in the form of fencing is possible and the costs can be afforded, the smallholder can also for example move towards the systems used under Traditional Free Range poultry production, in combination with grasing ruminants. The most common family poultry flock size that can be maintained by a family without special inputs in terms of additional feeding, housing and labour is up to about 20 birds. Small flocks of this size are able to scavenge sufficient feed in the household surroundings to survive and to reproduce.

Any attempt to increase flock size above these levels is likely to result in malnutrition and/or in a increase in the required daily foraging distances.

Fencing or confinement introduces the need to provide a supplemental feed in order to provide a balanced diet. In most cases this additional feed will need to be purchased.

Intensive Systems

These systems are used by medium to large-scale commercial enterprises, and are also used at the household level. Birds are fully confined either in houses or cages.

Capital outlay is higher and the birds are totally dependent on their owners for all their requirements; production however is higher. Intensive systems of rearing indigenous chickens commercially is uncommon. Instead, large and medium scale commercial production relies on the use of hybrid birds.

There are three types of intensive systems:

Deep litter system: Birds are fully confined within a house 3 to 4 birds/m^2) but can move around freely.

The floor is covered with a deep litter (5 to 10 cm deep layer) of grain husks (maize or rice), straw, wood shavings or a similarly absorbent, nontoxic material.

The fully enclosed system protects the birds from thieves and predators and is suitable for specially selected commercial breeds of egg or meat producing poultry (layers, breeder flocks and broilers).

Slatted floor system: Wire or wooden slatted floors are used instead of deep litter, which allow stocking rates to be increased to five birds/m^2 of floor space. Birds have reduced contact with faeces and are allowed some freedom of movement. Faeces can be collected from below the slatted floor and used as fertilizer.

Battery cage system: This is usually used for laying birds, which are kept throughout their productive life in small cages.

There is a high initial capital investment, and the system is mostly confined to large-scale commercial egg layer operations.

***Table** : Production and reproduction per hen per year under different management systems*

Production system	*Number of eggs per hen/year*	*Number of year-old chickens*	*Number of eggs for consumption & sale*	*Suitable breeds*
Scavenging (free-range)	20-30	2-3	0	local breeds
Improved scavenging and health care	40-60	4-8	10-20	local breeds
Semi-intensive	100	10-12	30-50	hybrids or local
Intensive (deep litter)	160-180	25-30	50-60	hybrids
Intensive (cages)	180-220	-	180-220	hybrids

Source: adapted from FAO (2004)

Use Common Sense

When constructing a poultry house and run, use common sense in designing the structure. Build the roof high enough and situate permanent structures such as nests, roosts, and feeding areas so that they are easy to access and that it is easy to clean all areas of the house. Where fully enclosed poultry houses are used, install doors so that they open inward. The door to an outside run can open outwards, so that when opening the door one does not push against and possibly damage any birds inside the run.

Use building materials which will be easy to clean and simple to disinfect when necessary.

The use of roofing felt, commonly found in small poultry houses available commercially in Europe, is not generally recommended as this tends to provide ideal conditions for the red mite. Slightly sloping the floor toward the door can help prevent flooding in the building and will make the building easier to spray out and dry between uses.

Provide shade: Site the house so that it provides maximum shade throughout the day.

Red Mite

Also called the Roost mite or Poultry mite, lives most of its life off the birds, sheltering in crevices and cracks in the poultry house. They can survive off the bird, without a blood feed, for up to six months and so can be extremely difficult to get rid of.

Mites are nocturnal feeders that hide during the day under manure, on roosts, and in cracks and crevices of the chicken house, where they deposit eggs. Populations develop rapidly during the warmer months and more slowly in cold weather. The life cycle may be completed in only 1 week.

One way of helping to prevent infestations is to empty the house, wash and disinfect it and then fill all obvious cracks that you can see and so reduce the mite's 'hiding places'.

Transmission is by dispersion of the mites through contact with infested birds, animals, or more commonly with inanimate objects. Transmission may also be via people moving between different farms.

Heavy infestations of either chicken mites or northern fowl mites decrease reproductive potential in males, egg production in females, and weight gain in young birds.

Paradigm and Visions: Network for Poultry Production and Health in Developing Countries

In spite of the fact that the majority of rural poor keep a small flock of chickens; poultry has long been neglected in the development community. Networks, based on rural poultry, have recently been established and the interest for using poultry as a means in poverty alleviation and food security programmes is increasing.

However, the accessibility to literature, documents, guidelines, manuals, etc. is a main constraint. Consequently, previous experiences are often lost and new projects or programmes often start from scratch. Even though the interest is increasing and more development professionals than ever before are involved in rural poultry keeping, ways of communication and sharing experiences are still in its conception phase.

A successful model has developed in Bangladesh in which the main elements are: community group formation, establishment of an enabling environment, and capacity building for establishing and maintaining a smallholder poultry sector. Till now, more than 1 million families have been established with poultry activities and within a period of five years the concept will probably be in operation in more than 50,000 villages. This development has been the inspiration both to formulate a paradigm and to emphasis more on visions and capacity building in the project formulation phase.

A paradigm in which experiences are accumulated and disseminated, a paradigm which learns from its own mistakes and successes and a paradigm which constitutes the framework for dissemination of experiences and information. With more than 1 billion people living in extreme poverty, it is stressed, that development programmes have to be designed with a build-in mechanism for replication, which mainly means local capacity buildings.

Network for Poultry Production and Health in Developing Countries

The success of a model developed in Bangladesh, in which poultry is used as a tool in poverty alleviation, fostered the idea to establish an institution with the objectives to develop methods, based on the same principle, to be used in poverty alleviation programmes in other developing countries.

A broad outline of the background and the intention of the Network were presented at this workshop in 1998 (Jensen 1998).

Poultry has in the past been, and still is, a neglected animal by the development community compared with other livestock. However, in the late 1970s the Bangladesh NGO BRAC identified poultry rearing as a source of income for the landless, particularly destitute women. Others have since con-firmed that poultry keeping is a common denominator for the majority of the poor in rural areas in developing countries. Currently the relationship between poor households and the poultry as the sole domesticated animal kept by the household is more or less recognised as a fact. Furthermore, about 70% of the rural landless women are directly or indirectly involved in poultry rearing activities, which there-fore represent skills known to them.

Paradigm

A paradigm means in this context:

1) a framework concept comprising a set of mutually supporting activities,
2) a set of values expressing or clarifying the impact of the framework activities on poverty alleviation, and
3) methods to continuously improve and disseminate the concept and knowledge related to the subject (capacity building).

The framework concept is based on a model developed in Bangladesh in close cooperation between the Department of Livestock Services (DLS) and the NGO Bangladesh Rural Advancement Committee (BRAC).

The basic feature of the model is a Smallholder with some 10 hens supported by a number of small entrepreneurs, all available in the village, to provide the inputs and the services needed to maintain a flock of 10 hens.

The concept is glued together by community groups, awareness programmes, training, and access to micro-credits. Even though the different entrepreneurs are established as an integrated production chain each unit operates on free market conditions and is free to sell to customers outside the integration chain.

The framework concept's main activities are: 1) establishing and maintaining community groups and activities in the community groups (the learning process), 2) establishing and maintaining an enabling environment – in this case: all the activities needed to establish and operate a small flock of poultry, and 3) local capacity building.

It is stressed that the concept is not a replication of the Bangladesh model, but adaptation of the principles to the prevailing structure and culture in a specific country.

The values are socioeconomic parametres, and not production parametres, and used in the process of adaptation in a specific country. Achievement made in single-discipline-oriented programmes is not sustainable even though the production efficiency is improved (Kitalyi 1998). Alam (1996) has in impact surveys focused more on socioeconomic parametres in order to document the positive impact of the smallholder model in Bangladesh.

By choosing socioeconomic parametres instead of production parametres the paradigm became circular rather than linear in its development. An activity, such as improved breed, may show a positive effect on the egg yield, but have a negative effect on the family livelihood because the new breed does not have the required brooding traits to produce and nurse chickens.

The values are the socioeconomic parametres with which to judge an increase in the families' livelihood security and thereby ensure a sustainable and circular development by continuously adjusting the mutually supporting activities, which constitute the framework concept.

The Methods used in poultry development programmes have till now been rather one-sided: vaccination campaigns or cockerel exchange programmes, and with nearly no feedback procedures in-build in the projects.

Terminologies used in village poultry are rather confusing and have often a different meaning. For some backyard poultry are the same as village poultry and for others not. For some indigenous breeds have a low productivity when the egg yield is below 50 eggs per year, for others 50 eggs per year have no meaning, but if the 50 eggs mean 4 clutches and 4 hatches and the outcome are 30 saleable chickens per year it is a remarkable and high productivity.

Accessibility to information and experiences is troublesome: with no textbooks, databases, journals, or other media in which results and findings are published. The main parts of publications are in proceedings known only to a limited number of development workers.

Different projects, based on rural poultry, are uncoordinated either because there are different donors involved or because different persons are responsible. The consequences are often that the same mistakes are passed over from one project to the next. The smallholder project in Bangladesh has, in this respect, been in a rather unique position because the implementing institutions in Bangladesh have been the same, DLS and BRAC, over 5 projects in more than a 10 year period and Danida, is or has been, involved in 4 of these projects as donor. The experiences from one project have been accumulated in the same institution, which are responsible for preparation and implementation of the succeeding project of same type.

With establishment of the International Network for Poultry Development supported by FAO and the Danish Network supported by Danida the first steps are taken to establishment of a learning process in which not only a terminology is developed, but also methods of adaptation and dissemination are refined.

The methods intended to be used by the Danish Network are:

(1) through a systematic feedback process to continuously improve the supporting concepts,

(2) adaptation, through pilot projects, of the framework concept to other countries than Bangladesh, and

(3) through capacity building make the concept and experiences available for wider application.

The capacity building is integrated in the pilot projects for local dissemination and in the Network for application in other countries and for diffusion of innovations.

Observations and Generalisations

The experiences with microcredits to the poor have over the past 2 decades clearly documented that the poor are credit worthy and that they are willing to take a calculated risk assuming they can comprehend the consequences (Todd 1996 and 1998). In the process of developing the microcredit programmes it has also been experienced that women are better managers of loans than men are. As BRAC expresses it: "BRAC's experience shows that as the poor rural women are con-strained to manage the entire household with extremely limited resources, they develop as better managers than their male counterparts. When a woman benefits, her entire household benefits and the impact is more sustainable".

When poor people, even illiterate, get the opportunity they behave rational: diversify the income generating activities, make savings, send children to schools, improve the family nourishment, and improve the family health. (Todd 1996 and Alam 1996). However, microcredit alone is not enough as expressed in a statement by the International Food Policy Research Institute at the Microcredit Summit Washing-ton January 31 1997:

'Rural finance alone will not relieve poverty. More credit does not necessarily mean less poverty. Appropriate policies and good governance are critical for creating an environment in which financial services can make a difference for the poor. People must be educated and healthy enough to use credit in productive activities. Efficient, functioning markets are also critical for small-scale farmers and entrepreneurs to obtain the inputs and outputs they need to produce and get their production market. Investment in social safety net, as well as roads, electricity and communication infrastructure, are necessary to enhance the impact of credit in relieving poverty.'

The activities constituting the framework concept are in line with this statement: the community groups are the educational part and the enabling environment enhances the impact of credit and alleviation of poverty. It is the local environment, in which the poor accidentally live that cause the poverty and not the poor that create the miserable environment. Furthermore, even illiterate people can work themselves out of poverty if they get an opportunity they can comprehend. Consequently, focus must be on establishing an enabling environment.

Poultry are a unique tool to reach the poor women with minimum disturbance of the patriarchal family pattern. Traditionally, poultry are women's domain and the income from poultry is in the hand of women; consequently an increased income from a small flock of hens is easier kept under women's responsibility than an increased income from cattle. In order to develop a standard concept with which to reach a vast number of the rural poor, especially the poor women, poultry are a unique tool because:

- The majority of the poor are familiar with poultry keeping.
- The investment is low.
- The turnover is fast.
- The results are visible.
- The educational values of using poultry in participatory learning process are considerable.

Based especially on experiences from Bangladesh it seems obvious that creating an environment in which the poor women have the opportunity to establish an income generating activity without any subsidy involved at user level most effectively does poverty alleviation. Poultry is a unique entity as a starting point for such a development and microcredit is an essential, but not the only tool, in creating an enabling environment.

Vision and Scale

Potential and policies: More than a billion people live in extreme poverty on less than $ 1 a day - and the pressing question is: how can development assistance be most effective at reducing global poverty?

The OECD's Development Assistance Committee (DAC) has as its task to be the principle strategy think-tank of the major bilateral donors. These strategies are elaborated on in its publication "Shaping the 21st century: 'the role of Development Cooperation'. Some of the goals set forth by the donor community are:

- Reducing by one half the proportion of people living in extreme poverty by 2015.
- Making progress towards equality of the sexes and the empowerment of women by eliminating disparities in primary and secondary education by 2005.
- Implementing national strategies for sustainable development in all countries by 2005 to ensure losses of environmental resources are reversed both nationally and globally by 2015.

> *"All this points to a different role for aid. Development assistance is more about supporting good institutions and policies than providing capital. Money is important, of course, but effective aid should bring a package of finance and ideas and one of the keys is finding the right combination of the two to address different situations and problems." (quoted from* Assessing Aid*)*

Eradication of poverty has priority on the development agenda and it is realised that money alone is not enough; new ideas and new concepts have to be developed and implemented. To reach the target set by DAC approximately 50 million people must every year be lifted above the poverty line, from now till 2015, and this in addition to the existing effort with which the proportion of poor only is maintained at status quo.

Furthermore, institutional development, with emphasis on health and education, must be improved.

According to FAO the absolute number of chronically undernourished people rose between 1990-2 and 1994-96 in three out of five developing regions of the world, namely Sub-Saharan Africa, Near East and North Africa, and South Asia. The number rose from 822 millions in 1990-92 to 828 millions in 1994-96. Only in East and South East Asia did the number decline while it stagnated in Latin America and the Caribbean.

To change the present status quo situation in reduction of poverty from approximately 20% of the world population living below the absolute poverty line to 10% by 2015 will require not only political visions, but also wider visions in our methods to eradicate poverty than currently practised.

Capacity building will be a key word in the effort to reach such a target. BRAC as well as the Grameen Bank has proved that it is possible to build up an institutional capacity with which to reach a large proportion of the poor. Both NGOs have each a member accession of about 200,000 per year. Assuming each member represent a family of 7 persons, then 2.8 million new persons are every year involved in a programme in which the main objective is poverty alleviation in a country like Bangladesh. In the case of BRAC, more than 50% of the new members start with poultry as their first income generating activity.

It is an unrealistic dream that capacity building in other countries can be done in the same way as in Bangladesh. However, if the political statements regarding poverty alleviation are to have any meaning, it is an absolute necessity that capacity building is integrated in poverty alleviation projects in such a way that a country-wide dissemination programme succeeds pilot projects in a specific country.

Interaction between Project Design and Visions

Between 1973 and 1986, the World Bank lent US$ 19 billion for nearly five hundred (498) rural development projects, the total cost of which were estimated at $ 50 billion. The outcome for these was a large proportion of failures, especially in sub-Saharan Africa (World Bank, 1988 passim). In the words of the Bank's own, commendably self-critical evaluation: 'the Bank apparently lost sight of the reality that the cost of failures, in what were identified from the outset as risky experiments, would be born by the borrower countries and not by the Bank'. The evaluation concludes that there are many lessons to be learnt. They included problems arising from:

- Institutional and managerial complexity.
- Lack of the viable technical packages (which had been assumed); and:
- Supply-driven lending, high targets, and urgent large-scale action without pilot projects.

The above project strategy certainly has had a vision, 500 projects of similar types and an investment of US$ 50 billion. However, it could have been interesting to see the results if the rural development programmes, from the very beginning, had had a framework concept in which experiences were transmitted from previous projects to new projects through a circular feedback process. Then we would today have had a project concept that had been developed and refined through 500 steps.

The interaction between the vision and the project-design may be the most important element in the design of the pilot project. The pilot project must encompass involvement of all the stakeholders intended to participate in the final project.

Human resource development and institutional capacity development for dissemination (replication) are important element in the Network strategies and will as such be an integral part of the project design in the different countries

International Network for Family Poultry Development: Origins, Activities, Objectives and Visions

The development of animal production requires cooperation in sharing information and experience within and between countries. In order to identify appropriate technologies that will improve the performance of locally available animal and feed resources within the rural system, there is a need to coordinate, codify, amplify and broadcast across state and language barriers the individual efforts undertaken at various locations. We must find, collate, share and crosscheck all procedures which "minimise risks and optimise production with low-cost inputs; conserve and improve the farming system resource base; minimise wastes and environmental degradation; and recycle wastes for animal feed or energy supply." (Qureshi, 1993). This requires extensive contacts and information sharing, such as can be found in a network.

Types of Network

Agricultural Research Networks (ARN)

An ARN is a group of individuals or institutions linked together because of a commitment to solve a common agricultural problem or set of problems and to use existing resources more effectively.

ARN can function at three levels: the simple exchange of ideas, methodologies and research results; scientific consultation between individuals or groups working on a common problem, conducting their research independently and sharing their results at common meetings; and collaborative research with joint planning and monitoring of a common research problem.

ARN generally seek to focus research efforts, based in institutions, on an agreed set of problems in such a way that the benefit anticipated by individual participants exceeds the cost that they incur and that the sum of benefits exceeds aggregate costs. This implies a high degree of organisation and formality in agreeing on over-all research agenda, research methods, allocation and scheduling of tasks, division of financial and other resources, format and manner of reporting.

Information Exchange Network (IEN)

An IEN is a group of individuals or institutions linked together on a voluntary basis with the primary objective of exchanging information on themes of professional interest in cost-effective ways. A key feature of IEN is their low cost. Once set up, network members can share information on their own experience, and benefit from that of others, at a lower resource cost to them than would be the case if they had to submit articles and take out subscriptions to professional journals.

IEN can function in a variety of ways. They can support several methods of information exchange. An obvious possibility is an element of research which can help to generate information to be exchanged, to bring network members together in active assignments and to focus attention on specific themes. IEN can use several methods of communication. Conventionally, these include newsletters and network papers, as well as workshops and symposia. In addition to written communication, farmer - to - farmer visits and electronic media can be very important in networking.

Organisations with a Networking Function (ONF)

Much information exchange in agricultural development is carried out by ONF, which are structured primarily around objectives other than networking. Examples of ONF include: Information providing organisation which promote high rates of member interaction by well-targeted mailing lists and question-and-answer services (e.g. Technical Centre for Agricultural and Rural Cooperation, CTA); Advocacy and activist networks (Genetic Resources Action International, GRAIN; Inter-national Federation of Organic Agriculture Movements, IFOAM); and Informal local groupings such as farmer and craft groups or cooperatives.

Research Partnerships by an ONF

The International Livestock Centre for Africa (ILCA) developed a strategy of partnerships with national and international institutions in collaborative research, training and information exchange. The mechanisms used included interinstitutional collaboration, informal consultations, information exchanges, joint planning and programming and collaboration in research, training and institution building. However, the primary focus was on collaborative research networking with the national agricultural research systems (NARS) in sub-Saharan Africa (Walsh, 1993). Before the reorganisation that resulted in the new International Livestock Research Institute (ILRI), there were a total of nine ILCA-associated networks; three commodity research networks and six discipline-related networks.

The need to provide access to relevant livestock information for scientists; to focus their attention on the rural farmstead where the majority of livestock is held and to provide opportunities for training and shared experiences on livestock in farming systems justifies the establishment of a research and development IEN.

Prerequisites for Networking

Networking in agricultural development has a strong element of voluntary collaboration and philanthropic ideals. The aim must be to improve living standards among those seeking to make a living under difficult farming conditions the complex, diverse and risk prone farming areas (Chambers et al., 1989). This cannot be a process in which those who are able to supply most resources (such as money and information) for a network are the ones who receive the most benefit from it. It must be a collaborative process working to support those who wish to develop solutions to difficult problems.

A prerequisite for networking is that there is self-motivation among the members. A group of individuals or organisations actively and consistently communicating or exchanging around a central theme is essential to an IEN. Regardless of the method of communication, it is a process of exchange.

This rules out a one-way information flow from one agricultural research centre simply seeking to inform an audience of its activities. Networks function better if the members feel a sense of ownership as in a voluntary, non-governmental organisation.

INFPD - A typical Information Exchange Network

The International Network for Family Poultry Development (INFPD) formerly African Network on Rural Poultry Development (ANRPD) is a typical IEN.

Its aims are to:

1. consolidate knowledge of rural family poultry production and coordinate efforts to develop it.
2. serve as a forum for exchange of ideas and resources, comparison of methods and evaluation of results.
3. document results and disseminate information.
4. coordinate training programmes and develop human resources. Identify research and development priorities, funding sources and cooperation opportunities.

Anrpd has contributed towards rural poultry development in Africa in the following areas.

1. identified human and institutional resources.
2. facilitated contact and mutual development.
3. guided emphasis onto rural poultry development.
4. made rural poultry work respectable.

Some of the problems we have encountered include:

1. too many dormant members.
2. insufficient face-to-face interaction.
3. incomplete or non-implementation of plans.
4. insufficient feedback. And
5. inadequate impact and coordination.

Origins—The Challenge at Hameln

It was at the 1987 Poultry Workshop in Hameln, Germany, that I heard Werner Bessei speak about rural poultry production. He was then the FAO Poultry Production Officer. Before his paper I had never thought of studying the production system of traditional poultry production and I was completely at a loss for words, data or insight to contribute.

I had lived with this system all my life but knew almost nothing about it even after about 15 years of poultry research experience. In response to this humiliation, I dedicated myself to learn all I could on this topic and to share such information with anyone interested. I asked Dr. Bessei to do a literature search of the available FAO database on rural poultry.

The Idea of a Network

The result was disappointing. It was then I decided to contact as many African scientists as possible to ask for all available publications, completed studies, raw data, research proposals and development ideas on rural poultry. But why should they supply these to me? What do they get in return?

Why, of course, they will get the fruits of my own search in the same area. So it became clear to me that what was needed was a network of scientists and development workers interested in rural poultry development. A proposal was presented to the 1989 Poultry Workshop at Hameln (Sonaiya, 1989).

FAO Midwifery

The network pregnancy was delivered with the assistance of the FAO's Professor Bessei who asked me to mount a workshop on rural poultry development in Africa in November 1989. It was at that workshop that participants agreed to become founding members of the African Network for Rural Poultry Development.

CTA Wet Nursing

It provided part funding for the proceedings of the 1989 workshop and in 1990, mounted its own international seminar on smallholder rural poultry production in Thessaloniki, Greece. This seminar reaffirmed the formation of ANRPD and further developed its aims and objectives. In addition, CTA funded the meeting of the steering committee for the next 3 years. The former Deputy Director, Dr. Werner Trietz and the former Chargé de Mission, M. Dominique Hounkonou, were very instrumental to this sup-port.

IDRC Step Fathering

The International Development Research Centre of Canada (IDRC) also part financed the publication of the proceedings of the 1989 workshop and was the first organisation to fund a proposal for collaborative research submitted by some net-work members in 1991. It has also provided funds for the publication of the 1997 ANRPD workshop. The Senior Programme Officer, Prof. Ola Smith has, from inception supported the Network.

From Rural Through Village to Family

But what is rural poultry? Is a turnkey, all-in all-out, 2 million broilers per cycle factory sited in a village qualified to be called rural poultry? What about small scavenging flocks kept in the periurban areas or even in the high-density residential areas? Rural Poultry was agreed by participants at the 1989 workshop in Ile-Ife, Nigeria, to be " any genetic stock of poultry (unimproved and/or improved) raised extensively or semi-intensively in relatively small numbers (less than 100 at any given time).

There is minimal investment on inputs with most of the inputs generated in the farmstead, labour is not salaried but drawn from the family with production geared essentially towards home consumption or savings." (Sonaiya, 1990).

If rural poultry is understood locationally, the term village poultry will be preferable. Evidently, poultry kept within the villages will be owned and managed by villages and as long as there are villages, there will be village poultry. The Australians and their colleagues in S.E Asia prefer this term. But then, village poultry is still locational and as Aini (1990, 1998) has pointed out, very large flocks are kept in the villages under a semi-scavenging system.

The current FAO Animal Production Officer responsible for small stock Rene Branckaert has solved this problem by suggesting that the focus of our network should be on family poultry. When due cognisance is taken of the clientele of our network in the developing countries of the southern hemisphere, the concept of family poultry can hardly be understood in the same way as a Dutch family man-aged poultry farm with 50,000 layers in automatically controlled aviaries.

In our reference areas of Africa, Asia and Latin America, family poultry refers to small flocks managed by individual farm families in order to obtain food security, income and gainful employment for women and children.

Past and Current Activities

As a result of two international symposia in Germany, one international workshop in Nigeria and an international seminar in Greece, the African Network for Rural Poultry Development (ANRPD) or Réseau Africain pour le Développement de l'Aviculture en millieu Rurale (RADAR) was established as a cost-effective means of coordinating efforts in research and development activities in Africa. At the meeting in Greece, nearly all participants from Africa indicated that Newcastle disease was the most significant disease of rural poultry.

Many contributors proposed the creation of an international network for the coordination of research on New castle disease in village poultry in Africa. It was however agreed to incorporate this effort into the more general ANRPD. At the last general meeting of the network held in 1997 at M'Bour, near Dakar, Senegal, it was agreed that the network be renamed the International Network for Family Poultry Development (INFPD) or Reseau International pour le Développement Aviculture Familiale (RIDAF) and its coverage be extended to Asia and Latin America.

The network now deals not only with "rural" but also with development of periurban, family-operated poultry production. It was agreed that all efforts be made to encourage network activities in Asia and Latin America and specialists were identified as correspondents for their regions: Asia Prof Aini Ideris (Malaysia), Dr. D.P. Singh (India); Latin America - Antonio Jose Solarte (Colombia) and Dr. Niels Kyvsgaard (Nicaragua and Den-mark). These specialists were identified to the FAO Representatives in their countries and regions and the membership and facilities of the former ANRPD were made available to them. Hence they became additional members of the INFPD Executive Committee.

The INFPD is an independent voluntary association targeted at researchers, policy makers, educationists, students and development workers (including NGOs) operating in or interested in Africa. By 1997, there were 369 members from 36 African, 9 European, 7 Asian and 3 Latin American countries. The network publishes a newsletter in English and French three times a year by email with a hard copy version once a year. Every two years, there is a meeting and general assembly.

An Executive Committee administers the Network with 8 members and 4 correspondents. This committee performs the following functions:

1. Search for funds.
2. Identify sub-regional and national coordinators.

3. Assemble and disseminate information on research grants, training pro-grammes, exchange visits, ect.
4. Encourage and facilitate interdisciplinary and international cooperation in research, development and training activities. And
5. Organise, every 2 years, meetings of Network members.

Research Activities

Three Priority Areas Were Identified at the Thessaloniki Meeting

1. Epidemiological assessment of the different strains of Newcastle disease virus and evaluation of the efficiency of traditional remedies, polyvalent vaccines and new vaccines with regard to the ease of field application, cost, integration into the feeding and management system and long-term environmental and public health implications.

   This priority is currently being addressed by the Coordinated Research Pro-gramme (CRP code: 313.D3.20.19) of the Joint FAO/IAEA Division on Nuclear Techniques in Food and Agriculture titled: "Assessment of the effectiveness of vaccination strategies against Newcastle disease and Gumboro disease using Immunoassay-based technologies for increasing Farmyard poultry production in Africa."
2. Evaluation of the genotypes and their relationship to productivity in the different species raised singly or in combination under the three types of smallholder system in different ecological zones. Of particular importance is the study of the ability of different poultry species and breeds to utilise high fibre and unconventional feeds for egg and meat production.

   Genotype evaluation is an on going preoccupation, which was given a boost by the 5 year EEC STD grant to a 5 nation team led by Prof. Dr. Peter Horst. The International Foundation for Science has also given individual research grants to Network members for genotype evaluation, NCD vaccine evaluation and management system evaluation. Similarly, the evaluation of unconventional feeds is proceeding in the laboratories of many members but the specific study of high fibre utilisation has not been sufficiently accomplished.
3. Socioeconomic analysis of the efficiency of the smallholder production system in relation to labour and animal productivity,

rural food security, stability of the rural labour force and potential contribution to family incomes (especially of women). Assessment of the sustainability of the various technological interventions within the three types of smallholder production systems.

This priority area is still outstanding for most countries and is sorely needed. The excellent analysis of the Bangladeshi situation can be a guide to other countries.

Development Activities

The following priorities have been identified:

1. Collection and compilation of unconventional feed stuff and assessment of their nutritive value.

Many members in different countries have addressed this and some compilations have been published (e.g. Sonaiya, 1995). A member has proposed a book on Poultry Feedstuffs in Africa.

2. Promotion of smallholder producers' associations and their linkage with foreign partner associations.

There are up to 10 associations and organisations catering to smallholder farmers listed in the 1997 ANRPD Directory. They include Farmers Development Union, Ibadan Nigeria; Groupement Ferme d'Action Maraichere de Kokoro, Bangui, Central Africa Republic; Salima Permaculture Group, Salima, Malawi, etc. There is a need to do more in promoting linkages between these associations.

3. Installation of a regional programme for training, research and development for poultry (and other small domestic stock).

The FAO is preparing a Manual on Family Poultry Production, which can be used in training and development work. The biannual meetings of the Network have a strong, though informal, training element, which is continued in the published proceedings as well as in the other publications of the Network. The Newsletter, which is in its 10th year, has provided information to about 1000 people annually. It is available to students and staff of institutions, colleges and departments that sub-scribe. Up to 30 such training related institutions subscribe. They include Kwa Zulu Natal Poultry Institute, South Africa; Chevalier Training Farm, Fiji; Africa University, Zimbabwe; Egerton University, Kenya; Bunda College, Malawi; Live-stock Training Institute, Tangeru, Tanzania; College of Agriculture, Ekpoma, Nigeria; and even mass media organisations such as the BBC, and the Farmers Radio Network, Toronto, Canada.

There is still a need for regional training programmes especially for Africa and Latin America but this will have to be packaged in a new form.

Visions for INFPD

As we approach the 21st Century, there is need to reassess the role and function of the INFPD. Will there still be a need for Family Poultry? What factors will govern the production of family poultry and its contribution to poverty alleviation and to promotion of gender equity?

Family Poultry of the Future

There is no doubt that urbanisation is increasing very rapidly in the developing countries. Therefore, a greater percentage of family poultry will be produced in periurban and even urban areas. At the same time, there is a drive in the developed countries away from intensive poultry production. We are aware now that development of family poultry must be in the direction of greater intensification of resource use be it feed, health, housing or management resources. According to Christensen (1998), "Producers have realised that adopting a simpler technology of production without giving up the production and disease control (both animal and public health) gains of the intensive poultry systems is not easy... Approaching the frontier from the intensive side carries the same pitfalls of compromising consumer expectations by indiscriminate use of therapeutic and prophylactic medicines rendered obsolete by modern poultry production systems, as does the approach from the subsistence side."

Less than 60 years ago, most families in Europe kept small flocks of laying hens in their backyards to provide animal protein at an affordable cost. Most developing countries are at that stage and have a lot to learn from the developed countries on how to develop a poultry industry. European countries have forgotten the effect of free-range poultry on families and their environments but can now observe such conditions in the developing countries. The INFPD can play a crucial role in this exchange of data and experience. The greater scientific, technological and financial powers of developed countries are needed to complement the wider spread in biodiversity and sociocultural circumstances of the developing countries.

Future Role of INFPD

For agrarian countries, livestock plays a strategic role in the farming system. Ruminants have been duly recognised in this capacity. However, poultry and other small stock are the livestock of the poor and the women. Approaches to livestock development for poverty alleviation and gender equity that have less negative impact on the environment must properly involve poultry. INFPD wishes to continue to play a role in the search, evaluation and institutionalisation of such approaches.

Hence, our priorities and visions for the future centre on institutionalisation of the family poultry paradigm for poverty alleviation and gender equity. It is in this light that our linkage with the World's Poultry Science Association (WPSA) should be seen. Through WPSA, we wish to place family poultry in the purview of the world's poultry scientists. The Executive Committee of WPSA meeting in Jerusalem agreed that a symposium be mounted during the World Poultry Congress (WPC) 2000 in Montreal, Canada. The theme is Family Poultry and Food Security. This is a result of the deliberate and consistent förderung of Dr. Rene Branckaert who has served as Chairman of our Advisory Committee for about 7 years. He ensured the inclusion of the 1st International Symposium on Rural Poultry Development Policy in the XIX WPC in Amsterdam, and the Rural Poultry Development Symposium in the XX WPC in New Delhi.

At the country level, individual INFPD members are encouraged to be active in the local branch of the WPSA and other animal science associations with the aim of focusing the attention of their peers on family poultry. Wherever possible, national networks on family poultry are encouraged. Tanzania and Kenya already have or are ready to launch their networks. Burkina Faso, Cameroon, Mali, Senegal and Togo should be able to initiate their own networks. In Nigeria, the partnership with WPSA seems to be flourishing. An annual Poultry fair including a symposium is the public vehicle chosen. The years ahead will witness a greater awareness and interest in family poultry research and development by our scientists, extension agencies and policy makers. Early this month (March 7-11, 1999), an NGO (FAcE-PaM) organised an international seminar around the theme: Promoting Sustainable Small scale Livestock production towards Reduction of Malnutrition and Poverty in Rural and Suburban Families in Nigeria. The Director of Federal Livestock Department, who attended, has directed that a national programme for rural poultry development be devised.

At the international level, information management will remain a priority. With the continued assistance of the FAO, electronic conferencing will be a major emphasis. We would like to see the provision of facilities for groups (based on region or specialisations) to exchange information, have group meetings and for the whole Network to treat various topics and subtopics. The results of such group activities and general conferences should be storable, searchable and readable by old and new members of the Network. Such information should be organised hierarchically in a relational manner so as to allow access by category, key word, etc. from our web site on the Internet.

Along with these reports, we would like to develop interactive courses of family poultry at different levels of competence. Some other courses should be useable for promotion of family poultry in primary and secondary schools as well as in a poultry advisory system. Taken together, these can represent the desired regional training pro-grammes when appropriate languages and local coordinators are used.

Electronic conferencing notwithstanding, INFPD will continue to promote regular face-to-face interaction of its members every 2 years. Those coinciding with the World Poultry Congress (WPC) will be held during WPC and those in the years between WPC will be held independently.

The Role of Women in Poultry Development: Proshika Experiences

Bangladesh has a population of 123 million and a density of 755 people per square km. The country is classified as a developing country. 86% of the population live in rural areas and 48.9 percent are women. Female-headed households constitute 9 per cent of all households and 30 per cent of all poor households. Although the vast majority of the rural population is poor and illiterate, the women are the poorest and have a much lower literacy rate than men. The problems affecting the economic and social status of women in Bangladesh are huge and complex.

Women in the Labour Force

The size of the civilian labour force in Bangladesh is 56 million (in 1996-97) of which 45.8 million are rural and 10.2 million are urban. The number of men that constitutes the labour force is 34.7 million while the number of women is 21.3 million. In the case of women, most of the productive activities are performed within the household. Though they may appear to be unemployed, generally they are overworked. Women make a direct contribution to the economy through their participation in agricultural and non-farm activities and indirectly, they contribute through their work in the household (Alamgir, 1997). In the household, women are responsible for all the domestic work such as cleaning, cooking, washing, rearing of children, raising poultry and vegetables, tending animals.

Women in Livestock Production in Bangladesh

Rural women traditionally play a very important role in raising livestock. In most cases, they are solely responsible for goats/sheep and poultry. They also take care of the health of the animals and birds. However, the household job they perform is unpaid and the traditional extension service does not make much contribution to raise their skills. The women are by passed by banks and other money lending institutions

The article states further that, "Nothing in the article shall prevent the state from making special provision in favour of women or children or for the advancement of any backward section of citizens."

Proshika, one of the largest private voluntary development organisation (PVDOs) in Bangladesh, has made a special provision for rural women to generate employment and income for them through its different development and employ-ment and income generating (EIG) programmes, especially poultry development activities.

Proshika: Goal, Objectives and Programmes

Proshika Manobik Unnayan Kendra, in short Proshika has been working on development activities for the poor since its establishment in 1976. The goal of Proshika is to promote sustainable development for poverty free, productive, environmentally sound, democratic and just Bangladesh. In order to achieve this goal Proshika's objectives are: i) structural poverty alleviation; ii) environmental protection and regeneration; iii) improvement in women's status; iv) increasing people's participation in public institutions and v) increasing peoples capacity to gain and exercise democratic and human rights.

In order to achieve these goals and objectives, Proshika has formulated certain strategies, which are implemented through the following programmes:

a) Organisation building among the poor
b) Development education
c) Promoting self-reliance through EIG activities: (i) A Livestock (ii) Irrigation (iii) Sericulture (iv) Apiculture and (v) Fishery Development Programme.
d) Environmental protection and regeneration: (i) Social Forestry Programme (ii) Ecological Agriculture

e) Universal education.
f) Health education and infrastructure building.
g) Gender Relations Co-ordinations Cell (GRCC)
h) Urban poor development.
i) Housing
j) Impact Monitoring & Evaluation Cell (IMEC)
k) Development Support Communication (DSCP)
l) Disaster preparedness and Management.
m) Institute for Development Policy Analysis and Advocacy (IDPAA).

Commercial Poultry Development Programme in Proshika

About 89 per cent of the rural households rear poultry. The poultry development programme is one of the largest components under the Livestock development programme in Proshika

Proshika works both with commercial poultry rearing and is involved in a semi-scavenging model through three collaborative projects in 55 thanas. The Proshika poultry development setup is as follows:

Training for Livestock Support Services and Poultry Rearing

Preference is to women to equip them to participate in vaccination, as paravets and as advanced paravets. Some women are trained as feed sellers and simultaneously to give extension service. A practical oriented training course is on "Poultry rearing and project management". It is for group members to provide them technical skills and knowledge about project planning and proper implementation processes.

Credit Support from a Revolving Loan Fund

Women groups are provided loans from a Revolving Loan Fund (RLF) for poultry production. The poultry projects have been categorised into 5 types: a) Broiler rearing, b) Chick rearing, c) Layer farming, d) Cockerel rearing, e) Duck rearing.

Development of Parent Stock Farm & Hatcheries

Proshika has developed a demonstration farm in the Human Resource Develop-ment Centre (HRDC) at Koitta, The demonstration projects include: mini scale dairy, poultry hatchery with a commercial parent stock farm (broiler & layer), fodder production, biogas, etc. This hatchery is producing 9000 broiler chicks and 2000 layer chicks weekly.

A modern controlled poultry hatchery has been operating since 1998 with the capacity of 30000 broiler chicks weekly. Another poultry hatchery in Payrabondh at Rangpur is under construction at the time of writing and two modern poultry hatcheries with commercial parent stock farms are planned. The objectives of these farms are to ensure the supply of Day Old Chicks (DOCs) to the broiler and layer project holders.

Livestock Compensation Fund

Proshika developed a livestock and poultry compensation fund to compensate losses due to death of animals and poultry. The Fund has minimised the risk of financial losses due to projects undertaken by the people.

Technical Staff

The technical extension services are provided through trained livestock and veterinary graduates, who are deployed in all the Area Development Centres as technical workers. They provide technical support, instructions, advice and act as facilitators in skill development training courses.

Chapter 2

Poultry Keeping

Poultry

Definition of Layer Chicken

Layer chicken is a special species of hen for egg production which have to raise from when they are one day old. They start laying egg from 18-19 weeks of age. They remain laying egg continually to their 72-78 weeks of age. They produce about one kg of egg eating 2.25 kg of food during their laying egg period. For the purpose of producing hybrid layer egg, considering the desired characteristics mating different types of cock and hen trough long research making more egg laying hens are called hybrid layer.

Name of Layer Species

According to the nature and colour of egg, layer hens are of two types.

Brown Egg Hen

Brown egg hen are relatively large in shape. They eat more food. Lays big egg than other species. Egg shell is brown. There are many types of brown hen. Such as, Isa Brown, Hi Sex Brown, Sever 579, Lehman Brown, Hi Line Brown, Bab Cock BV-300, Gold Line, Bablona Tetro, Bablona Harko, Havard Brown etc.

Layer Hen Selection

You have to keep in mind some essential information before selecting the layer for your farm. You have to select those species which is suitable for your business.

- For layer production you have to chose high productive hen correctly.
- All types of hen does not produce equal eggs.
- Well quality hen have to collect.
- If the desired characteristic has a fame then layer can be collect.
- Layer should have to collect from a famous hatchery.

White Egg Hen

This types of hen are comparatively small in size, relatively eat less food and the colour of egg shell is white. Isa White, Lehman White, Nikchik, Bab Cock BV-300, Havard White, Hi Sex White, Sever White, Hi line White, Bovanch White etc. are good example of white egg laying hen.

Figure : *Layer Chicken*

Cockerel Keeping

During the first weeks of their birth many cockerel suffers by not to drink water due to caring one place to another. So adequate drink water system have to make in their brooder house and they have to be trained to drink water. They can easily get energy if you mix 5% glucose with water. Any types of high quality multivitamin can be served mixing with water, suggested by electrolyte production companies instruction. Multivitamin and electrolyte are very useful while farmer carry cockerel from a long distance. It reduces tiredness and lack of water and make the cockerel normal.

Vaccination and it's Importance

Vaccination program is a must for cockerel to keep them free from disease. There are some advantage of vaccination.

- It make illness resistance power.
- Keep the hen free from infective illness.
- Disease prevalence will be less.
- Mortality rate will be in a moderate stage.

There are many types of vaccine available for layer hen. Marex, Ranikheth, Gamboro, Bruchaities, Bosonto, Salmonela, Karaiza etc. Those vaccine are used for layer chicken.

Marex, Ranikheth, Gamboro are used for broiler chicken.

Before Vaccination

You have to maintain some rule before vaccination.

- Hen should Catch very carefully.
- Chicken have to vaccinate without any strain.
- There is no need to vaccinate the ill hen.
- Vaccination equipment must have to be boiled in boiling water.
- Vaccination program should happen when the weather is cold.
- Preventive vaccine is always applicable to healthy bird.

Growing Cockerel Keeping Method

You have to maintain the suggestion listed below for growing layer cockerel.

- They have to provide special care until they reach 4-5 weeks of age.
- After brooding good quality pullet have to serve.
- It result good in future.
- It produce egg highly.
- Make the chicken healthy.
- Increase body weight.
- So it is very important to select and keeping of pullet.

Egg Production from Commercial Layer Farm

Egg production from a commercial layer farm depends on farm management. If the farmer takes special care of their farm then the production will high.

- 5% of hen lays egg when they reach the age 20 weeks.
- About 10% lays at the age of 21 weeks.
- When they reach 26-30 weeks they produce highly. It may be different according to their strain.
- After laying maximum number of eggs they remain paused for a few days.
- And after that the egg production reduces slowly.
- The size of egg big according with their increasing of laying egg rate.
- The hen grows until they are 40 weeks old.
- Weight of egg increase until they reach 50 weeks of age.

Method and Importance of Lip Cutting

- To reduce fight among them.
- To prevent food loss.
- Cockerel should cut their lip at the age of 8-10.
- Growing cockerel should 8-12 weeks.
- For cockerel 0.2 cm from their nose should cut.
- Growing hen 0.45 cm should cut.
- Both lips have to cut.
- Both lips should cut off one after another.
- Block chick triaming machine, hi speed triaming machine and dibikar can use to cut the lip.

When Lip Should Not Cut:

- Two days after or before vaccinate.
- After or before serving medicine like sulfur.
- If the hen in a strain.
- Much weather change.
- If the hen start laying egg.

Water mixed with vitamin "K" have to served three days before cutting lips. Lips cutting instrument should washed with antiseptic.

Test the edge and temperature of blade. You have to be careful so that the eyes and tongue are not damage. Choose cold weather to cut the lip. Lip cutting process should observed by an experienced technician. After cutting lip water should served in a deep pot. Extra meat should provide.

Figure: *Hen*

Food

There are many company in our country who make food for layer hen. You can buy food from local market. You have to be sure that the food you bought are enriched with value. Protein and mineral are very important for layer production. However, you also can make layer food by your own.

- 2% of calcium should provide for two weeks before they are 2 weeks old.
- If the weight is not expected you can serve starter feed up to eight weeks.
- Food can served two or three time in a day to next 18 week.
- Demand of food increase fast when they start laying.
- Food should provide according to the age and weight.
- Food should not less when they are laying if their weight increase.

Poultry Farming

Poultry farming is the practice of raising domesticated birds such as chickens, turkeys, ducks, and geese, as a subcategory of animal husbandry, for the purpose of farming meat or eggs for food.

More than 50 billion chickens are raised annually as a source of food, for both their meat and their eggs. Chickens raised for meat are called broilers, whilst those raised for eggs are called laying hens. In total, the UK alone consumes over 29 million eggs per day. Some hens can produce over 300 eggs a year. Chickens will naturally live for 6 or more years. After 12 months, the hen's productivity will start to decline. This is when most commercial laying hens are slaughtered.

The majority of poultry are raised using intensive farming techniques. According to the Worldwatch Institute, 74 percent of the world's poultry meat, and 68 percent of eggs are produced this way. One alternative to intensive poultry farming is free range farming.

Friction between these two main methods has led to long term issues of ethical consumerism. Opponents of intensive farming argue that it harms the environment and creates health risks, as well as abusing the animals themselves. Advocates of intensive farming say that their highly efficient systems save land and food resources due to increased productivity, stating that the animals are looked after in state-of-the-art environmentally controlled facilities. A few countries have banned cage system housing, including Sweden and Switzerland. Consumers can still purchase lower cost eggs from other countries' intensive poultry farms.

Techniques

Free-range

Free-range poultry farming consists of poultry permitted to roam freely instead of being contained in any manner. In the UK, the Department for Environment, Food and Rural Affairs says that a free range chicken must have daytime access to open-air runs during at least half of its life. Unlike in the United States, this definition also applies to eggs. The European Union regulates marketing standards for egg farming which specifies a minimum condition for Free Range Eggs states that "hens have continuous daytime access to open-air runs, except in the case of temporary restrictions imposed by veterinary authorities". In free-range broiler systems, the chickens are given continuous access to an outdoor range during the daytime and sheds where they are housed at night. Free-range chickens grow more slowly than intensive chickens. They live at least 56 days. In the EU, each chicken must have one square metre of outdoor space.

Figure: *Free range chickens being fed outdoors*

Free-range poultry production requires that the poultry have access to the outside. In some cases this means the poultry are raised on pasture, enabling the poultry to move around, forage for their natural diet and live in cleaner conditions than those in batteries. In some farms, the manure from free-range poultry can be used to benefit crops.

The benefits are also an increased growth rate and opportunities for natural behaviour such as pecking, scratching, foraging and exercise outdoors, as well as fresh air and daylight. Because they grow slower and have opportunities for exercise, free-range chickens have better leg and heart health and a much higher quality of life.

Finding suitable land with adequate drainage to minimise worms and coccidial oocysts, suitable protection from prevailing winds, good ventilation, access and protection from predators can be difficult. Excess heat, cold or damp can have a harmful effect on the animals and their productivity. Unlike battery farms, free range farmers have little control over the food their animals come across, which can lead to unreliable productivity.

Some free range farming in the UK, which accounts for 26% of production, has also come under criticism concerning animal welfare. This is due to some large-scale free range farms where social abnormalities arise due to having large numbers of birds in an outdoor space. Beak trimming due to cannibalism and infighting is common in this form of poultry farming as well as in batteries. Diseases are common and the animals are vulnerable to predators. In South-East Asia, a lack of disease control in free range farming has been associated with outbreaks of Avian influenza.

In organic systems, chickens are also free-range. Organic chickens are slower growing, more traditional breeds and live typically for around 81 days. They grow at half the rate of intensive chickens. They have a larger space allowance outside (at least 2 square metres and sometimes up to 10 square metres per bird).

Yarding

Figure : *Ducks and other poultry*

While often confused with free-range farming, yarding is actually a separate method of poultry culture by which chickens and cows are raised together. The distinction is that free-range poultry are either totally unfenced, or the fence is so distant that it has little influence on their freedom of movement. Yarding is common technique used by small farms in the Northeastern US.

Daily releases out of hutches or coops allows for instinctual nature for the chickens with protections from predators. The hens usually lay eggs either on the ground of the coop or in baskets if provided by the farmer. This technique can be complicated if used with roosters though, mostly because of difficulty getting them into the coop and to clean the coop while it is inside. This territorial nature is apparent while outside in which they have a brood of hens and sometimes even informal land claims. This can endanger people unaware of the existence of the territories who are attacked by the larger birds.

Intensive chicken farming

Figure : *Egg-laying chickens in battery cages*

Figure : *Egg-laying chicken 5 days out of battery cage*

In egg-producing farms, birds are typically housed in rows of battery cages. Environmental conditions are automatically controlled, including light duration, which mimics summer daylength. This stimulates the birds to continue to lay eggs all year round. Normally, significant egg production only occurs in the warmer months. Critics argue that year-round egg production stresses the birds more than normal seasonal production.

***Figure** : Broilers in a production house*

Meat chickens, commonly called broilers, are floor-raised on litter such as wood shavings or rice hulls, indoors in climate-controlled housing.

Poultry producers routinely use nationally approved medications, such as antibiotics, in feed or drinking water, to treat disease or to prevent disease outbreaks arising from overcrowded or unsanitary conditions. In the U.S., the national organisation overseeing chicken production is the Food and Drug Administration (F.D.A.). Some F.D.A.-approved medications are also approved for improved feed utilisation.

In egg-producing farms, cages allow for more birds per unit area, and this allows for greater productivity and lower space and food costs, with more efforts put into egg-laying.

In the U.S., for example, the current recommendation by the United Egg Producers is 67 to 86 in^2 (430 to 560 cm^2) per bird, which is about 9 inches by 9 inches. Modern poultry farming is very efficient and allows meat and eggs to be available to the consumer in all seasons at a lower cost than free range production, and the poultry have no exposure to predators.

The cage environment of egg producing does not permit birds to roam. The closeness of chickens to one another frequently causes cannibalism. Cannibalism is controlled by de-beaking (removing a portion of the bird's beak with a hot blade so the bird cannot effectively peck).

However, de-beaking does not fully prevent cannibalism it just reduces the damage. Most battery chickens are missing 30-70% of plumage by the time that they are spent.

Another condition that can occur in prolific egg laying breeds is osteoporosis. This is caused from year-round rather than seasonal egg production, and results in chickens whose legs cannot support them and so can no longer walk. During egg production, large amounts of calcium are transferred from bones to create eggshell. Although dietary calcium levels are adequate, absorption of dietary calcium is not always sufficient, given the intensity of production, to fully replenish bone calcium.

Under intensive farming methods, a meat chicken will live less than six weeks before slaughter. This is half the time it would take traditionally. This compares with free-range chickens which will usually be slaughtered at 8 weeks, and organic ones at around 12 weeks.

In intensive broiler sheds, the air can become highly polluted with ammonia from the droppings.

This can damage the chickens' eyes and respiratory systems and can cause painful burns on their legs (called hock burns) and feet. Chickens bred for fast growth have a high rate of leg deformities because they cannot support their increased body weight. Because they cannot move easily, the chickens are not able to adjust their environment to avoid heat, cold or dirt as they would in natural conditions. The added weight and overcrowding also puts a strain on their hearts and lungs. In the U.K., up to 19 million chickens die in their sheds from heart failure each year.

Indoor with Higher Welfare

Chickens are kept indoors but with more space (around 12 to 14 birds per square metre). They have a richer environment for example with natural light or straw bales that encourage foraging and perching. The chickens grow more slowly and live for up to two weeks longer than intensively farmed birds. The benefits of higher welfare indoor systems are the reduced growth rate, less crowding and more opportunities for natural behaviour.

Issues with Poultry Farming

Humane Treatment

Figure : *Chickens transported in a truck.*

Animal welfare groups have frequently criticized the poultry industry for engaging in practices which they believe to be inhumane. Many animal rights advocates object to killing chickens for food, the "factory farm conditions" under which they are raised, methods of transport, and slaughter.

Compassion Over Killing and other groups have repeatedly conducted undercover investigations at chicken farms and slaughterhouses which they allege confirm their claims of cruelty.

Conditions in intensive chicken farms may be unsanitary, allowing the proliferation of diseases such as salmonella and E. coli.

Chickens may be raised in total darkness; hens are most often kept in crowded wire battery cages with space less than that of a sheet of paper per hen, as opposed to cage-free or free range.

Rough handling and crowded transport during various weather conditions and the failure of existing stunning systems to render the birds unconscious before slaughter have also been cited as welfare concerns. Another animal welfare concern is the use of selective breeding to create heavy, large-breasted birds, which can lead to crippling leg disorders and heart failure for some of the birds. Concerns have been raised that companies growing single varieties of birds for eggs or meat are increasing their susceptibility to disease.

A common practice among hatcheries is the culling of newly born male chicks of egg laying breeds, since they don't lay eggs, and do not grow fast enough to be profitable for meat.

Debeaking

Laying hens are routinely de-beaked when young to prevent fighting and feather pecking. Animal rights activist claim this is bad because beaks are sensitive, and the usual practice of trimming them without anaesthesia is considered inhumane by some.

De-beaked chickens will peck much less than chickens with beaks, which animal behaviourist Temple Grandin attributes to guarding against pain. The chicken industry says that de-beaking is not painful. Others argue that the procedure causes lifelong chronic pain and discomfort and decreased ability to eat or drink.

Intelligence

Some groups which advocate for more humane treatment of chickens claim that chickens are intelligent. Dr. Chris Evans of Macquarie University claims that their range of 20 calls, problem solving skills, use of representational signalling, and the ability to recognise each other by facial features demonstrate the intelligence of chickens.

Antibiotics

Antibiotics have been used on poultry in large quantities since the 1940s, when it was found that the by products of antibiotic production, fed because the antibiotic-producing mould had a high level of vitamin B_{12} after the antibiotics were removed, produced higher growth than could be accounted for by the vitamin B_{12} alone.

Eventually it was discovered that the trace amounts of antibiotics remaining in the by products accounted for this growth.

The mechanism is apparently the adjustment of intestinal flora, favouring "good" bacteria while suppressing "bad" bacteria, and thus the goal of antibiotics as a growth promoter is the same as for probiotics. Because the antibiotics used are not absorbed by the gut, they do not put antibiotics into the meat or eggs. Antibiotics are used routinely in poultry for this reason, and also to prevent and treat disease.

Many contend that this puts humans at risk as bacterial strains develop stronger and stronger resistances. Critics point out that, after six decades of heavy agricultural use of antibiotics, opponents of antibiotics must still make arguments about theoretical risks, since actual examples are hard to come by. Those antibiotic-resistant strains of human diseases whose origin is known originated in hospitals rather than farms.

A proposed bill in the United States Congress would make the use of antibiotics in animal feed legal only for therapeutic (rather than preventative) use, but it has not been passed. However, this may present the risk of slaughtered chickens harbouring pathogenic bacteria and passing them on to humans that consume them.

In October 2000, the U.S. Food and Drug Administration (FDA) discovered that two antibiotics were no longer effective in treating diseases found in factory-farmed chickens; one antibiotic was swiftly pulled from the market, but the other, Baytril, was not. Bayer, the company which produced it, contested the claim and as a result, Baytril remained in use until July 2005.

To prevent any residues of antibiotics in chicken meat, any given antibiotics are required to have a "withdrawal" period before they can be slaughtered. Samples of poultry at slaughter are randomly tested by the FSIS, and shows a very low percentage of residue violations

Arsenic

Chicken feed can also include Roxarsone, an antimicrobial drug that also promotes growth. Roxarsone was used as a broiler starter by about 70% of the broiler growers between 1995 to 2000. The drug has generated controversy because it contains arsenic, which is highly toxic to humans.

This arsenic could be transmitted through runoff from the poultry yards. A 2004 study by the U.S. magazine Consumer Reports reported "no detectable arsenic in our samples of muscle" but found "A few of our chicken-liver samples has an amount that according to EPA standards could cause neurological problems in a child who ate 2 ounces of cooked liver per week or in an adult who ate 5.5 ounces per week." The U.S. Food and Drug Administration (FDA), however, is the organisation responsible for the regulation of foods in America, and all samples tested were "far less than the amount allowed in a food product."

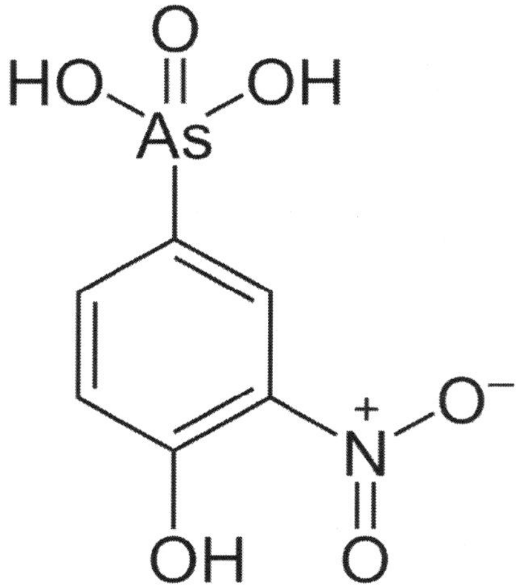

Roxarsone, a controversial arsenic compound used as a nutritional supplement for chickens.

Growth Hormones

Hormone use in poultry production is illegal in the United States. Similarly, no chicken meat for sale in Australia is fed hormones. Several scientific studies have documented the fact that chickens grow rapidly because they are bred to do so, not because of growth hormones. A small producer of natural and organic chickens confirmed this assumption:

"If this were 1948, you might have something to worry about. Using hormones to boost egg production was a brief fad in the Forties, but was abandoned because it didn't work. Using hormones to produce soft-meated roasters was used to some extent in the Forties and Fifties, but the increased growth rates of broilers made the practice irrelevant—the broilers got as big as anyone wanted them to get when they were still young enough to be soft-meated without chemicals.

The only hormone that was ever used in any quantity on poultry (DES) was banned in 1959, after everyone but a few die-hard farmers had given them up as a silly idea. Hormones are now illegal in poultry and eggs. The people who advertise "No hormones" are either woefully ignorant or are indulging in cynical fear-mongering, maybe both."

E. Coli

According to Consumer Reports, "1.1 million or more Americans are sickened each year by undercooked, tainted chicken." A USDA study discovered *E. coli* in 99% of supermarket chicken, the result of chicken butchering not being a sterile process. However, the same study also cautions that the type of *E. coli* turned up was in every case a non-lethal form distinct from the more dangerous "O157:H7" strain.

Many of these chickens, furthermore, had relatively low levels of contamination. Feces tend to leak from the carcass until the evisceration stage, and the evisceration stage itself gives an opportunity for the interior of the carcass to receive intestinal bacteria. (So does the skin of the carcass, but the skin presents a better barrier to bacteria and reaches higher temperatures during cooking).

Before 1950, this was contained largely by not eviscerating the carcass at the time of butchering, deferring this until the time of retail sale or in the home.

This gave the intestinal bacteria less opportunity to colonise the edible meat. The development of the "ready-to-cook broiler" in the 1950s added convenience while introducing risk, under the assumption that end-to-end refrigeration and thorough cooking would provide adequate protection. *E. coli* can be killed by proper cooking times, but there is still some risk associated with it, and its near-ubiquity in commercially farmed chicken is troubling to some. Irradiation has been proposed as a means of sterilising chicken meat after butchering.

Avian Influenza

There is also a risk that crowded conditions in chicken farms will allow avian influenza (bird flu) to spread quickly. A United Nations press release states: "Governments, local authorities and international agencies need to take a greatly increased role in combating the role of factory-farming, commerce in live poultry, and wildlife markets which provide ideal conditions for the virus to spread and mutate into a more dangerous form..."

Efficiency

Farming of chickens on an industrial scale relies largely on high protein feeds derived from soybeans; in the European Union the soybean dominates the protein supply for animal feed, and the poultry industry is the largest consumer of such feed. Two kilograms of grain must be fed to poultry to produce 1 kg of weight gain. However, for every gram of protein consumed, chickens yield only 0.33 g of edible protein.

Economic Factors

Changes in commodity prices for poultry feed have a direct effect on the cost of doing business in the poultry industry. For instance, a significant rise in the price of corn in the United States can put significant economic pressure on large industrial chicken farming operations.

World Chicken Population

The Food and Agriculture Organisation of the United Nations estimated that in 2002 there were nearly sixteen billion chickens in the world, counting a total population of 15,853,900,000. The figures from the *Global Livestock Production and Health Atlas* for 2004 were as follows:

1. China (3,860,000,000)
2. United States (1,970,000,000)
3. Indonesia (1,200,000,000)
4. Brazil (1,100,000,000)
5. Mexico (540,000,000)
6. India (495,000,000)
7. Russia (340,000,000)
8. Japan (286,000,000)
9. Iran (280,000,000)
10. Turkey (250,000,000)
11. Bangladesh (172,630,000)
12. Nigeria (143,500,000)

Poultry

Poultry is a category of domesticated birds kept by humans for the purpose of collecting their eggs, or raising for their meat and/or feathers.

These most typically are members of the superorder Galloanserae (fowl), especially the order Galliformes (which includes chickens, quails and turkeys) and the family Anatidae (in order Anseriformes), commonly known as "waterfowl" (e.g. domestic ducks and domestic geese).

Poultry also includes other birds which are killed for their meat, such as pigeons or doves or birds considered to be game, like pheasants. Poultry comes from the French/Norman word, poule, itself derived from the Latin word Pullus, which means small animal.

Poultry is the second most widely eaten meat in the world, accounting for about 30% of meat production worldwide, after pork at 38%.

Cuts of Poultry

Figure : *Cuts from a plucked chicken*

Figure : *The Poultry-dealer, after Cesare Vecellio*

The meatiest parts of a bird are the flight muscles on its chest, called breast meat, and the walking muscles on the first and second segments of its legs, called the thigh and drumstick, respectively. The wings are also eaten, usually (in the United States) without separating them, as in Buffalo wings; the first and second segment of the wings are referred to as drumette (meatier) and flat when these need to be distinguished, though these are technical terms. In Japan, the wing is frequently separated, and these parts are referred to as Kb½nCQ (*teba-moto* "wing base") and Kb½nHQ (*teba-saki* "wing tip").

Dark meat, which avian myologists refer to as "red muscle," is used for sustained activity—chiefly walking, in the case of a chicken. The dark colour comes from the protein myoglobin, which plays a key role in oxygen uptake within cells. White muscle, in contrast, is suitable only for short, ineffectual bursts of activity such as, for chickens, flying. Thus the chicken's leg and thigh meat are dark while its breast meat (which makes up the primary flight muscles) is white. Other birds with breast muscle more suitable for sustained flight, such as ducks and geese, have red muscle (and therefore dark meat) throughout.

Health

Consumption of large quantities of meat, including poultry, like overconsumption of any caloric food, has certain adverse effects which can include: obesity, heart disease, and constipation. In recent years, health concerns have been raised about the consumption of meat increasing the risk of cancer. Bird and animal fat, particularly from ruminants, tends to have a higher percentage of saturated fat vs. monounsaturated and polyunsaturated fat when compared to vegetable fats, with the exception of some tropical plant fats; consumption of which has been correlated with various health problems. The saturated fat found in meat has been associated with significantly raised risks of colon cancer, although evidence suggests that risks of prostate cancer are unrelated to animal fat consumption. USDA claims that consumption of meat as a source of protein in the human diet is crucial have been resoundingly contradicted by recent studies.

The correlation of meat, including poultry, consumption to increased risk of heart disease is controversial. A survey conducted in 1960 of 25,153 California Seventh-Day Adventists found that the risk of heart disease is three times greater for 45-64 year old men who eat meat daily, versus those who did not eat meat. In another study in 2010 involving over one million people who ate meat found that only processed meat had an adverse risk in relation to coronary heart disease. The study suggests that eating 50g (less than 2oz) of processed meat per day increases risk of coronary heart disease by 42%, and diabetes by 19%.

Chicken meat contains about two to three times as much polyunsaturated fat than most types of red meat when measured as weight percentage.

A recent study by the Translational Genomics Research Institute showed that nearly half (47%) percent of the meat and poultry in U.S. grocery stores were contaminated with *Staphylococcus aureus*, with more than half (52%) of those bacteria resistant to antibiotics.

Chicken (Food) Chicken the most Common Type of Poultry in the World

History

The modern chicken is a descendant of Red Junglefowl hybrids along with the Grey Junglefowl first raised thousands of years ago in the northern parts of the Indian subcontinent.

Chicken as a meat has been depicted in Babylonian carvings from around 600 BC. Chicken was one of the most common meats available in the Middle Ages. It was widely believed to be easily digested and considered to be one of the most neutral foodstuff. It was eaten over most of the Eastern hemisphere and a number of different kinds of chicken such as capons, pullets and hens were eaten. It was one of the basic ingredients in the so-called white dish, a stew usually consisting of chicken and fried onions cooked in milk and seasoned with spices and sugar.

Chicken consumption in the US increased during World War II due to a shortage of beef and pork. In Europe, consumption of chicken overtook that of beef and veal in 1996, linked to consumer awareness of Bovine spongiform encephalopathy or B.S.E.

Breeding

Modern varieties of chicken such as the Cornish Cross, are bred specifically for meat production, with an emphasis placed on the ratio of feed to meat produced by the animal. The most common breeds of chicken consumed in the US are Cornish and White Rock.

Chickens raised specifically for food are called broilers. In the United States, broilers are typically butchered at a young age. Modern Cornish Cross hybrids, for example, are butchered as early as 8 weeks for fryers and 12 weeks for roasting birds.

Capons (castrated cocks) produce more and fattier meat. For this reason, they are considered a delicacy and were particularly popular in the Middle Ages.

Edible Components

Figure : *Chicken in a public market*

Main:

- Breast: These are white meat and are relatively dry.
- Leg: Also called the "drumstick", this is dark meat and is the lower part of the leg.
- Thigh: Also dark meat, this is the upper part of the leg.
- Wing: Comprises three segments, one, shaped like a small drumstick, the middle segment, containing two bones, and the tip, sometimes discarded. Wings are often served as a light meal or bar food. Buffalo wings are a typical example.

Exotic:

- Chicken feet: These contain relatively little meat, and are eaten mainly for the skin and cartilage. Although considered exotic in Western cuisine, the feet are common fare in other cuisines, especially in the Caribbean and China.
- Head: Considered a delicacy in China, the head is split down the middle, and the brains and other tissue is eaten.

- Neck: This is served in various Asian dishes.
- Oysters: Located on the back, near the thigh, these small, round pieces of dark meat are often considered to be a delicacy.
- Pygostyle (chicken's buttocks) and testicles: These are commonly eaten in East Asia and some parts of South East Asia.

By-products:

- Carcase: After the removal of the flesh, this is used for soup stock.
- Chicken eggs
- Heart and gizzard
- Liver: This is the largest organ of the chicken, and is used in such dishes as Pâté and chopped liver.
- Schmaltz: This is produced by rendering the fat, and is used in various dishes.

Health Issues

Chicken meat contains about two to three times as much polyunsaturated fat than most types of red meat when measured as weight percentage.

Chicken generally includes low fat in the meat itself (castrated roosters excluded), however it is highly concentrated on its skin, which should be avoided when low intake of fat is necessary.

Cooking method is highly related to health issues related to chicken, with steam cooking with skin removed being considered one of the healthiest, and fried with trans fats one of the worst.

However according to a 2006 Harvard School of Public Health study of 135,000 people, people who ate grilled skinless chicken 5 or more times a week had a 52 percent higher chance of developing bladder cancer compared to people who didn't. However, such strong associations were not found in individuals regularly consuming chicken with skin intact. A recent study by the Translational Genomics Research Institute showed that nearly half (47%) percent of the meat and poultry in US grocery stores were contaminated with *S. aureus*, with more than half (52%) of those bacteria resistant to antibiotics.

Marketing and Sales

Juvenile chickens, of less than 28 days of age at slaughter in the United Kingdom are marketed as poussin. Mature chicken is sold as small, medium or large.

Figure : *Oven roasted chicken with potatoes.*

Whole mature chickens are marketed in the United States as fryers, broilers, and roasters. Fryers are the smallest size (2.5-4 lbs dressed for sale), and the most common, as chicken reach this size quickly (about 7 weeks).

Most dismembered packaged chicken would be sold whole as fryers. Broilers are larger than fryers. They are typically sold whole. Roasters, or roasting hens, are the largest chickens commonly sold (3-5 months and 6-8 lbs) and are typically more expensive. Even larger and older chickens are called stewing chickens but these are no longer usually found commercially.

The names reflect the most appropriate cooking method for the surface area to volume ratio. As the size increases, the volume (which determines how much heat must enter the bird for it to be cooked) increases faster than the surface area (which determines how fast heat can enter the bird). For a fast method of cooking, such as frying, a small bird is appropriate: frying a large piece of chicken results in the inside being undercooked when the outside is ready.

Chicken is also sold in dismembered pieces. Pieces may include quarters, or fourths of the chicken. A chicken is typically cut into two leg quarters and two breast quarters. Each quarter contains two of the commonly available pieces of chicken. A leg quarter contains the thigh, drumstick and a portion of the back; a leg has the back portion removed. A breast quarter contains the breast, wing and portion of the back; a breast has the back portion and wing removed. Pieces may be sold in packages of all of the same pieces, or in combination packages. Whole chicken cut up refers to either the entire bird cut into 8 individual pieces. (8-piece cut); or sometimes without the back.

A 9-piece cut (usually for fast food restaurants) has the tip of the breast cut off before splitting. Pick of the Chicken, or similar titles, refers to a package with only some of the chicken pieces. Typically the breasts, thighs, and legs without wings or back. Thighs and breasts are sold boneless and/or skinless. Dark meat (legs, drumsticks and thighs) pieces are typically cheaper than white meat pieces (breast, wings). Chicken livers and/or gizzards are commonly available packaged separately. Other parts of the chicken, such as the neck, feet, combs, etc. are not widely available except in countries where they are in demand, or in cities that cater to ethnic groups who favour these parts.

There are many fast food restaurant chains on both a national and global scale that sell exclusively or primarily in poultry products including KFC (global), Red Rooster (Australia), Hector Chicken (Belgium) and CFC (Indonesia). Most of the products on the menu in such eateries are fried or breaded and are served with french fries.

Cooking

Figure : *Marination of chicken for grilling.*

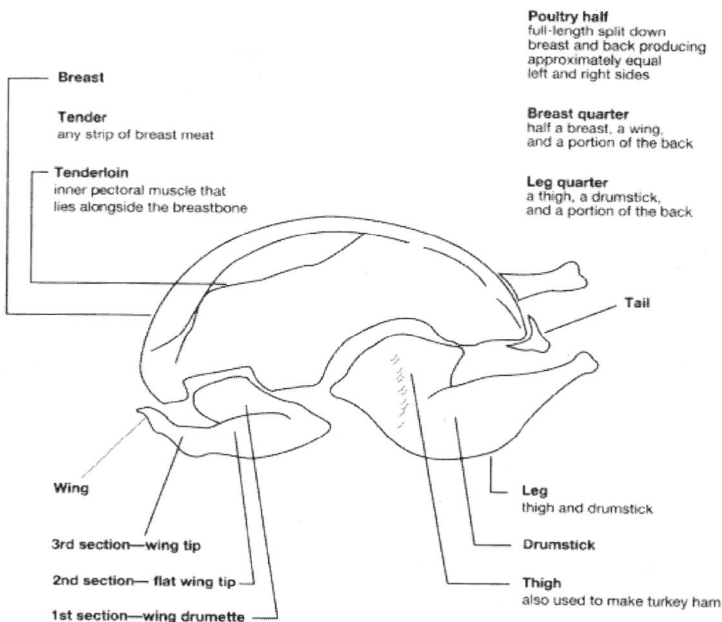

Figure : *The USDA classifies cuts of poultry in a manner similar to beef.*

Raw chicken can be frozen for up to two years without significant changes in flavour or texture. Chicken is typically eaten cooked as when raw it often contains *Salmonella*. Raw or rare chicken dishes appear in Ethiopian cuisine and Japanese cuisine.

Chicken can be cooked in many ways. It can be made into sausages, skewered, put in salads, grilled, breaded and deep-fried, or used in various curries. There is significant variation in cooking methods amongst cultures. Historically common methods include roasting, baking, broasting, and frying. Today, chickens are frequently cooked by deep frying and prepared as fast foods such as fried chicken, chicken nuggets, chicken lollipops or buffalo wings. They are also often grilled for salads or tacos.

Chickens often come with labels such as "roaster", which suggest a method of cooking based on the type of chicken. While these labels are only suggestions, ones labelled for stew often do not do well when cooked with other methods.

Some chicken breast cuts and processed chicken breast products include the moniker "with Rib Meat." This is a misnomer, as it is the small piece of white meat that overlays the scapula, and is removed with the breast meat. The breast is cut from the chicken and sold as a solid cut, while the leftover breast and true rib meat is stripped from the bone through mechanical separation for use in chicken franks, for example. Breast meat is often sliced thinly and marketed as chicken slices, an easy filling for sandwiches.

Often, the tenderloin (pectoralis minor) is marketed separately from the breast (pectoralis major). In the US, "tenders" can be either tenderloins or strips cut from the breast. Chicken bones are hazardous to health as they tend to break into sharp splinters when eaten, but they can be simmered with vegetables and herbs for hours or even days to make chicken stock.

In Asian countries it is possible to buy bones alone as they are very popular for making chicken soups, which are said to be healthy. In Australia the rib cages and backs of chickens after the other cuts have been removed are frequently sold cheaply in supermarket delicatessen sections as either "chicken frames" or "chicken carcasses" and are purchased for soup or stock purposes.

Freezing

Raw chicken maintains its quality longer in the freezer as compared to when having been cooked because moisture is lost during cooking. There is little change in nutrient value of chicken during freezer storage. For optimal quality, however, a maximal storage time in the freezer of 12 months is recommended for uncooked whole chicken, 9 months for uncooked chicken parts, 3 to 4 months for uncooked chicken giblets, and 4 months for cooked chicken.

Freezing doesn't usually cause colour changes in poultry, but the bones and the meat near them can become dark. This bone darkening results when pigment seeps through the porous bones of young poultry into the surrounding tissues when the poultry meat is frozen and thawed. It is safe to freeze chicken directly in its original packaging, however this type of wrap is permeable to air and quality may diminish over time.

Therefore, for prolonged storage, it is recommended to overwrap these packages. It is recommended to freeze unopened vacuum packages as is. If a package has accidentally been torn or has opened while food is in the freezer, the food is still safe to use, but it is still recommended to overwrap or rewrap it.

Chicken should be from other foods, so if they begin to thaw, their juices won't drip onto other foods. If previously frozen chicken is purchased at a retail store, it can be refrozen if it has been handled properly. Chicken can be cooked or reheated from the frozen state, but it will take approximately one and a half times as long to cook, and any wrapping or absorbent paper should be discarded.

Use of Roxarsone in Chicken Production

In many factory farms, chickens are routinely administered with the feed additive Roxarsone, a relatively benign organoarsenic compound which partially decomposes into inorganic arsenic compounds in the flesh of chickens, and in their feces, which are often used as a fertilizer. The compound is used to control stomach pathogens and promote growth.

A *Consumer Reports* study in 2004 reported finding "no detectable arsenic in our samples of muscle" but found "A few of our chicken-liver samples has an amount that according to EPA standards could cause neurological problems in a child who ate 2 ounces of cooked liver per week or in an adult who ate 5.5 ounces per week." However, the amounts found in these livers averaged to 460 part per billion; an amount still less than the 2,000 parts per billion limit set by the FDA. The FDA has found there to be no significant impact for the use of Roxarsone in the production of chicken, turkey or swine.

Chickens as Pets

Chickens can be tamed by hand feeding and simply by being handled.

Figure : *Baby chickens, dyed unnatural colours, sold as pets at a Market in Oaxaca, Mexico.*

Some people are afraid that roosters will become aggressive, but this problem can easily be avoided if the rooster is handled right. Breeds such as Silkies and many bantams are generally docile, making them ideal pets for owners with small children. Some cities in the United States allow chickens as pets but others ban them. Some may only ban roosters due to the crowing.

City ordinances, zoning regulations or health boards may determine whether chickens may be kept. A general requirement is that the birds be confined to the owner's property, not allowed to roam freely. There may be restrictions on the size of the property or how far from human dwellings a coop may be located, etc.

The so called "urban hen movement" harks back to the days when chicken keeping was much more common, and involves the keeping of small groups of hens in areas where they may not be expected, such as closely populated cities and suburban areas. In Asia, chickens with striking plumage have long been kept for ornamental purposes, including feather-footed varieties such as the Cochin and the Silkie from China, and the extremely long-tailed Phoenix from Japan. Asian ornamental varieties were imported into the United States and Great Britain in the late 1800s. Distinctive American varieties of chickens have been developed from these Asian breeds. Poultry fanciers began keeping these ornamental birds for exhibition, a practice that continues today. Individuals in rural communities commonly keep chickens for both ornamental and practical value.

HousingA chicken coop is a housing where chickens are kept. Inside there will often be nest boxes for egg laying along with perches on which the birds can sleep. Backyard coops are small and fenced, often with chicken wire, allowing chickens an area to roam, peck and hunt insects. Chicken tractors are floorless coops which can be dragged about a yard. Some backyard chickens are allowed to free range, sleeping in coops. Urban chicken keeping has led to manufactured chicken coops such as the Eglu, which are designed for tight spaces and have a tidy look. Chicken waterers and feeders are an important part of keeping chickens as pets. There are hanging waterers/feeders, nipple waterers and nipple cup waterers.

Show Chickens

Chicken shows may be found both at county fairs and sanctioned shows hosted by regional poultry clubs. 4,000 or more birds may be entered in some shows. The Poultry Club of Great Britain sanctions poultry shows in Britain while the American Poultry Association and the American Bantam Association do likewise in America. Such organisations also work with poultry breed clubs of people who have interests in specific breeds of chickens.

Free-range Eggs

Free-range eggs are eggs produced using birds that are permitted to roam freely within a farmyard, a shed or a chicken coop. This is different from factory-farmed birds that are typically enclosed in battery cages. The term "free-range" may be used differently depending on the country and its laws.

Legal definition or even nonexistent depending on the country. For example, the U.S. Department of Agriculture requires only that the bird spends part of its time outside, and allows egg producers to freely label these eggs as free-range.

Many producers will label their eggs as *cage-free* in addition to or instead of *free-range*. Recently, US egg labels have expanded to include the term "barn-roaming," to more accurately describe the source of those eggs that are laid by chickens who do not range freely but are confined to a barn instead of a more restrictive cage.

Figure : *Free-range hens.*

Cage-free egg production includes barns, free-range and organic systems. In the UK, free-range systems are the most popular of the non-cage alternatives, accounting for around 28% of all eggs, compared to 4% in barns and 6% organic. In free-range systems, hens are housed to a similar standard as the barn or aviary. In addition, they have constant daytime access to an outside range with vegetation. In the EU each hen must have at least 4 square metres of space.

Non-cagegunjan systems may be single or multi-tier (up to four levels), with or without outdoor access. Indoor non-cage systems are also referred to as aviaries (for systems with multiple tiers) or barn systems. The Laying Hens Directive stipulates that from 1 January 2007 (1 January 2002 for newly-built or rebuilt systems), non-cage systems must provide the following:

- A maximum stocking density of 9 birds/m^2 of "usable" space (units in production on or before 3 August 1999 may continue with a stocking density up to 12 birds/m^2 until 31 December 2011)
- If more than one level is used, a height of at least 45 cm between the levels
- One nest for every seven hens (or 1m2 of nest space for every 120 hens if group nests are used)
- Litter (e.g., wood shavings) covering at least one-third of the floor surface, providing at least 250 cm^2 of littered area per hen
- 15 cm of perching space per hen.

In addition to these requirements, free-range systems must also provide the following:

- One hectare of outdoor range for every 2500 hens (equivalent to 4 m^2 per hen; at least 2.5 m^2 per hen must be available at any one time if rotation of the outdoor range is practised)
- Continuous access during the day to this open-air range, which must be "mainly covered with vegetation"
- Several popholes extending along the entire length of the building, providing at least 2 m of opening for every 1000 hens.

Case studies of free-range systems for laying hens across the EU, carried out by Compassion in World Farming, demonstrate how breed choice and preventive management practices can enable farmers to successfully use non beak-trimmed birds.

Cost

Based on data in the European Commission's socioeconomic report, it costs €0.66 to produce 12 battery eggs, €0.82 to produce 12 barn eggs and €0.98 to produce 12 free-range eggs. So 12 free-range eggs cost €0.32 more to produce than 12 battery eggs, and 12 barn eggs cost €0.16 more to produce than 12 battery eggs. This means that one free-range egg costs 2.6 Eurocents more to produce than a battery egg, and a barn egg costs 1.3 Eurocents more to produce than a battery egg. The Commission's report concludes that, if costs were to increase by 20%, which it says is the type of percentage increase in terms of variable costs that producers are likely to face as a result of switching to free-range, the industry will potentially suffer a loss of producer surplus of €354 million (EU-25). This appears to be a substantial sum.

If, however, this increased cost were borne not by farmers but by consumers paying a little extra for eggs, each EU citizen would only have to pay less than €1 extra per year, as the human population of the EU-25 is around 460 million.

The margins achieved by producers for barn and free-range eggs are appreciably higher than those available for battery eggs. The Commission's socioeconomic report shows that margins for free-range eggs are around twice as high as those for battery eggs.

Retailers

A number of major retailers already have an express policy of only selling free-range eggs or of not selling battery eggs. Some retailers apply this policy not just to shell eggs but also to eggs used in baked goods and processed products such as ready-made meals, quiches, and ice cream. In the UK, The Co-Operative and Marks & Spencer sells only free-range shell eggs and uses only free-range eggs in their entire range of baked goods, processed products, and ready-made meals. Waitrose sells only non-cage shell eggs, and uses only free-range eggs in their processed products and ready-made meals.

As of 1 January 2007 (with one minor exception), all Austrian supermarkets no longer sell battery eggs. Many retailers in the Netherlands, including Albert Heijn and Schuitema (subsidiaries of Ahold), Laurus (including Edah, Konmar and Super de Boer), Dirk van den Broek (including Bas van der Heijden and Digros), Aldi and Lidl sell only free-range shell eggs. Three Belgian supermarkets: Makro, Colruyt and Lidl, no longer sell battery eggs. The Commission's report states that Sweden's move away from conventional battery cages has been aided by the decision by the four largest retailers (who between them account for 98-99% of the Swedish retail market) to stop stocking conventional battery eggs.

Misconceptions

Free range does not imply in any way that the hens were fed any differently than on normal commercial farms. The label "free roaming" does not describe feed supplies, which means that free-range hens can be fed the same animal-derived by products or GMO crops as in other non-organic farms. This is also the main reason why free-range eggs are cheaper than organic eggs.

Consumers of free-range eggs want eggs from hens that are kept under traditional low-density, free-range conditions.

Critics of EU-style free-range regulations point out that commercial free-range egg farming, in general, does not live up to these consumer requirements, since the regulations allow the use of yarding rather than free range.

Yarding combines a high-density poultry house with an attached fenced yard, and both its methods and results are closer to high-density confinement than true free range.

Free-range eggs may be broader, and have more of an orange colour to their yolks due to the abundance of greens and insects in the diet of the birds. An orange yolk is, however, no guarantee that an egg was produced by a free-range hen. Feed additives such as marigold petal meal, dried algae, or alfalfa meal can be used to colour the yolks.

Nutritional Content

Studies suggest the nutritional content of eggs from genuine free-range hens (hens that forage daily on a grass range) is superior to that of eggs produced by conventional means.

These studies report higher levels of Omega 3 and Vitamins A and E, and lower levels of total fat, saturated fat, cholesterol, and Omega 6.

An egg study comparing free-range eggs to the U.S. Department of Agriculture (USDA) nutrient data for commercial eggs. The findings showed that free-range chicken eggs produced the following results:

- 1/3 less cholesterol
- 1/4 less saturated fat
- 2/3 more vitamin A
- 2 times more omega-3 fatty acids
- 3 times more vitamin E
- 7 times more beta-carotene.

A study by the U.S. Department of Agriculture in 2010 using industry standard measures of shell strength, the height of egg whites (Haugh unit) and their protein and crude fat content determined that there were no nutritional benefits to free-range eggs when compared to factory eggs. However the study did not measure the types of fat in the eggs nor did it measure differences in vitamin and essential fatty acid content.

Organic Egg Production

Differences between "Free Range" and "Organic"

Significant differences cover feed, medication, and animal welfare. Organic hens are fed organic feed; it is prohibited to feed animal by products or GMO crops - which is not disallowed in free range environments; no antibiotics allowed except in emergencies (in free range, it is up to the farmer, but the same levels of antibiotics as conventional farming is allowed); required animal welfare standards in organic farms, which can improve the quality of both the eggs and the meat - low stress levels lead to superior quality of animal products, a fact long known and used in the production of the famous Wagyu beef.

Look for labels that say both "Organic" and "Free-Range". Cage free only means that they are not in a cage, but are usually still confined in barns at high population densities.

Living Conditions

Figure : *At an organic farm (in Bruthen, Victoria) chickens sometimes end up laying eggs somewhere other than the owner expected*

Organic Feed

Organic feed is grown by certified organic farmers. To become a certified organic farmer, the crop must be free of genetically modified organisms (GMOs). The crop must be free of GMOs and synthetic fertilization for three years before it can be certified for organic usage. If the crop is contaminated by cross-fertilization, the crop is rendered useless for organic grading. Finally, there can be no animal by-products fed to the poultry.

In the United States, "organic" egg production means that the flock may not live in cages and must have access to the outdoors. Organic egg producers in the United States do not always grant meaningful outdoor access to their organic laying hens; most industrial-scale organic egg producers generally build small wood or concrete porches attached to the henhouses, which passes as "outdoor access."

Antibiotics

Organic egg producers cannot feed low-level antibiotics to the poultry. Antibiotics are only allowed during an outbreak of infection or disease.

Moulting

Some farms induce molting in their flocks to affect egg production. In organic egg farms, the birds are allowed to go into a natural moult but are not induced.

Animal Welfare

Similar to all other forms of egg production in the United States, organic production is also regulated by animal welfare audit system. Mistreatment of the chickens could potentially lead a farmer to losing their organic certification. Thus some of the arguments from animal activist organisations that egg production is cruel and inhumane do not necessarily apply to organic raised hens. On the other hand, male chicks who are born on organic or free-range egg farms are still discarded, by the use of lethal gas, because they do not produce eggs. The American Organic Standards still allow beak trimming as a means to lessen injury to the birds.

Pastured Poultry

Pastured poultry :is a sustainable agriculture technique that calls for the raising of laying chickens, meat chickens (broilers), and/or turkeys on pasture, as opposed to indoor confinement. Humane treatment, the perceived health benefits of pastured poultry, in addition to superior texture and flavour, are causing an increase in demand for such products. Joel Salatin of Swoope, Virginia, helped to reintroduce the technique at Polyface Farm, and wrote his book *Pastured Poultry Profits* to spread the idea to other farmers. Andy Lee and Herman Beck-Chenoweth expanded on Salatin's techniques, and created some of their own.

The American Pastured Poultry Producers' Association (APPPA) was formed to promote pastured poultry. Its membership consists largely of pastured poultry farmers. Though pasture feeding improves the nutritive quality of ruminant meats, the effect of pasture feeding on poultry meat composition is not well established. One trial showed low impact of pasture feeding on vitamin E and fatty acid composition.

The pens that house the fowl can be made from wood and scrap metal or out of PVC pipe and white tarps. Pastured poultry is also gaining popularity because it helps the farmer, through reducing capital costs, and increasing pasture fertility. It is very well suited for incorporation within a system of managed intensive grasing. Pastured Poultry is not limited to chickens and turkeys. It includes a variety of other birds, including ducks, geese and exotics in the poultry family.

Free-Range Poultry

Figure : *A chicken tractor in use as part of a pastured poultry system*

Herman Beck-Chenoweth reintroduced the free-range system that was the most popular way to raise poultry in the U.S. from the 1930s through the 1960s. The system allows birds to range freely during the day and be safely sequestered on secure skid houses over night. The addition of a guard animal, such as an Komondor or Anotolian Shepperd dog, controls predators. In the Modern American Free-Range Poultry Production System birds are much less crowded and freer to practice normal bird behaviour than in any other pasture based system. Although frequently listed as a "pasture" method, free-"range" refers to the length of the forage. Cows graze "pasture" which is forage over six inches long. "Range" refers to short forage of 2-4 inches. Free-Range is a very sustainable production system that improves the farmer's soil and produces poultry with strong bones and meat with good "mouthfeel". Combined with proper aging after slaughter the meat is tender and flavourful.

Eggmobiles

As part of their pastured poultry process, some farmers use a mobile house that houses egg laying hens. This is called an eggmobile. Popularised by Joel Salatin, this model has been adapted across the United States. The concept is to allow chickens to move two to three days behind cows, which are rotationally grasing. The chickens scratch through cow manure, harvesting their own larvae and insects, spreading out manure and helping to reduce the fly population on the cows. The eggmobile contains nesting boxes that the hens use to lay eggs. The nesting boxes can be made of many materials, such as this design at Nature's Harmony Farm in Georgia which uses milk crates.

The foraging helps the hens to produce eggs with an "orangish" yolk and a thicker albumen. This is owed to the beta carotene that the grass provides to the eggs. A test done by Mother Earth News revealed that eggs laid by hens foraging on pasture have shown:

- 1/3 less cholesterol
- 1/4 less saturated fat
- 2/3 more vitamin A
- 2 times more omega-3 fatty acids
- 3 times more vitamin E
- 7 times more beta carotene.

National Chicken Council

The National Chicken Council (NCC), based in Washington, D.C., is the non-profit trade association representing the United States chicken industry. NCC is a full-service trade association that promotes and protects the interests of the chicken industry and is the industry's voice before the United States Congress and United States federal agencies. NCC member companies include chicken producer/processors, poultry distributors, and allied industry firms. The producer/processors account for approximately 95 percent of the chickens produced in the United States. The NCC's website contains a video about biosecurity in the poultry industry and links to its own avian influenza page, and additional references regarding avian influenza on its "About the Industry" page. The industry also is sponsoring a separate website with news, information, and resource links.

Modern agriculture and farming : Welcome to online farming guide. Bangladesh is an agricultural country. The economy of our country is mostly depend of agriculture. A major part of our population are directly or indirectly involved with agriculture. Income source of our people are limited. So they raise, poultry, ducks, dairy, fish etc. to earn some extra income. So, in this site we have tried to show you some possible way of earning. Not only for Bangladesh but also about Indian agriculture and almost all Asian agricultural countries.

We have launched in the concept of modern agriculture. We don't like traditional agriculture. Modern agriculture has changed the total agricultural process. So modern agricultural revolution is a must. To be success we need the proper, correct and experimental information based on modern technology. Which can make our dream true fast. Farmers to our country are doing their farming by the knowledge and experience which they obtain from their relatives. Modern technology in agriculture sector did not make any progress. Many eligible and famous researchers in our country are constantly researching and getting good results. But the main issue is that their results are not being published well. In some cases, the results of the research are restricted to journals and publication. We want to share the achievement of our researchers to the mass people. There are many agricultural university in our country to improve our agricultural condition. Many types of agricultural journals available in every parts of the worlds. Modern meditation, ideas, technology in agriculture and seeds in the land, to ensure the maximum use of fertilizer, raising poultry, ducks, dairy, goats, fish and many more, we have tried our best to collect all about this in our site. Due to improve the socioeconomic condition of our people we encourage land based farming as well as cattle, birds and the fish farming. Not only the well production, we have to be ensure about the market of our products.

Chapter 3

Poultry Brooding

Avian Incubation

Incubation is the process by which birds hatch their eggs, and to the development of the embryo within the egg. The most vital factor of incubation is the constant temperature required for its development over a specific period. Especially in domestic fowl, the act of sitting on eggs to incubate them is called brooding. The action or behavioural tendency to sit on a clutch of eggs is also called broody, and most egg-laying breeds of poultry have had this behaviour selectively bred out of them to increase production.

In most species, body heat from the brooding parent provides the constant temperature, though several groups, notably the Megapodes, instead use heat generated from rotting vegetable material, effectively creating a giant compost heap while Crab Plovers make partial use of heat from the sun.

The Namaqua Sandgrouse of the deserts of southern Africa, needing to keep its eggs cool during the heat of the day, stands over them drooping its wings to shade them. The humidity is also critical, and if the air is too dry the egg will lose too much water to the atmosphere, which can make hatching difficult or impossible. As incubation proceeds, an egg will normally become lighter, and the air space within the egg will normally become larger, owing to evaporation from the egg.

In the species that incubate, the work is divided differently between the sexes. Possibly the most common pattern is that the female does all the incubation, as in the Coscoroba Swan and the Indian Robin, or most of it, as is typical of falcons. In some species, such as the Whooping Crane, the male and the female take turns incubating the egg. In others, such as the cassowaries, only the male incubates.

The male Mountain Plover incubates the female's first clutch, but if she lays a second, she incubates it herself. In Hoatzins, some birds (mostly males) help their parents incubate later broods.

Incubation times range from 11 days (some small passerines and the Black-billed and Yellow-billed Cuckoos) to 85 days (the Wandering Albatross and the Brown Kiwi).

In these latter, the incubation is interrupted; the longest uninterrupted period is 64 to 67 days in the Emperor Penguin. It can be an energetically demanding process, with adult albatrosses losing as much as 83 g of body weight a day.

Some species begin incubation with the first egg, causing the young to hatch at different times; others begin after laying the second egg, so that the third chick will be smaller and more vulnerable to food shortages. Some start to incubate after the last egg of the clutch, causing the young to hatch simultaneously.

Incubation Periods

Bird	*Incubation Period (days)*
chicken	20–22
duck	26–28
finch	11–14
goose	25–28
ostrich	35-45
parrot	17–31
pheasant	24
pigeon	10–18
quail	21–23
swan	33–36
turkey	28

Veterinary and Poultry Farm Formulations

Our Veterinary and Poultry Farm Formulations is a unique formulation consisting of minerals, proteins, vitamins live yeast culture, which are intended to keep the cattle and poultry healthy. Further, these formulations are made up of pure and effective constituents, which are in right composition and thus, do not pose any side effects.

Our clients can avail these formulations in customised packaging at very reasonable prices in market.

Utilisation of Poultry Feed Resources by Smallholders in the Villages of Developing Countries

If land is arable, and production from it is sustainable, it should be used to produce food for people. The products of land suitable for neither cultivation nor for tree crops, and the by products of crop production and processing, are then available for animal production for the benefit of the human population, including nutrition. Ruminants can utilise all animal feeds, but monogastric species such as poultry and pigs utilise some feeds more efficiently.

The monogastrics do not make efficient use of high fibre diets, or diets which are low in essential amino acids and vitamins. Whereas ruminants can handle a large bulk of fibrous feed, and the organisms in the rumen break down some of the fibre as well as producing the essential amino acids and some vitamins. In some circumstances ruminants utilise high quality feed as a supplement to roughage as efficiently as for the production of meat and/or eggs by monogastrics (Preston & Leng, 1987). Superficially the rational distribution of animal feed resources seems obvious; but in practice the choice may not be clear-cut. There are cultural aversions to some species, the need for draught power may make large ruminants essential, and the preferences and purchasing power of consumers will influence production. Moreover, for the poorer families, trading of livestock and their products is often driven by family need and social obligations, rather than by market forces.

Feed Resources

Possible feed resources include:

- Household waste, including the waste from households which do not keep chickens.
- Materials from the environment including:
  - *metazoans such as worms, snails and insects. There is a conception that the village environment is a cornucopia of high quality feed - worms, snails, insects, green pick and seeds. In fact a village with a significant chicken population is stripped bare, an observation which is loudly endorsed by folk who do not keep chickens; but who wish to keep a kitchen garden.*

   - *grain products from cultivating, harvesting and processing*
   - *green pick*
   - *seeds*
- Cultivated and wild fodder materials:
   - *grasses, herbs, and fodder trees - grased, or cut and carried*
   - *water plants - Lemna, azolla, duck weed, Ipomoea aquatica*
- Nontraditional feed materials.

When the shortage of feed for scavenging chickens, particularly the lack of high quality protein, is considered, there is a tendency to invoke the prospects for unconventional or nontraditional feeds. Such feeds fall into two categories.

There are the by products from local industries, some of which are listed above, most of which are already fully utilised in one way or another. There is also a group of potential sources of animal protein such as cultured snails, earth-worms, termites, frogs and unicellular protein and vegetable protein, particularly from water plants. Expertise, labour and capital are necessary to produce the potential sources of animal protein so they are beyond the reach of village people. There is hope for vegetable protein in particular environments. There is no large untapped source of nutritional wealth out there; but there may be some jam for the bread.

- By products from local industries such as palm and tree crops, fishing, fish and crustacean culture, animal slaughtering, fibre (cotton seed, kapok seed), rubber seed, fruit processing and brewing
- By products from larger industrial units
- Prepared commercial feed
- Imported crop by products
- Imported feed grains and legumes.

Clearly all of the feed resources listed above are not available in every environment, or in every developing country. They are listed in a rough order of the ease with which village families have access to them. Thus village families have absolute control over their household waste, limited control over materials from the environment, and access at a price to the by products of local small industrial units; but virtually no access to the by products of larger industrial units. Some developing countries have problems with the supply and quality of feed for their intensive poultry industry. This is largely due to shortage of foreign exchange for importation of raw materials, and large-scale producers may well bid up the price of smaller scale resources to fill the gap. At the same time there are feed resources being exported to developed countries and one would hope that they would be utilised at source in the future.

How much feed Wealth do you have?

If a chicken project is contemplated for a district or country, or one is already in progress, it is wise to determine in advance, the size of the feed resource base, which in turn determines the potential biomass of the project flock. It is not difficult, mainly government statistics and some simple calculations. Just answer the following questions.

The feed resource base:

- What is the inter-regional trade in animal feed? Exports (+) and imports (-).
- Will the project area drain feed resources from other regions and thus preclude development there? (-)
- What is the distribution of demand for feed resources between commodities - chickens, milk, pigs, small ruminants, large ruminants and others? (-)
- What is the supply of by product feed resources?

If the amount of each major crop (rice, wheat, corn, soy, mustard, rape, —?) in the catchment area is known, then the gross amount of each by product can be calculated. Some crops may be processed locally and some may be sent out of the region (or imported into it) for processing.

There may also be industrial by products (fish processing, abattoirs, and prawn farming—?)

The total by product can then be separated into that which is:

– retained by the grower (-)

– exported (-)

– available in the market (+)

The rates of growth/decline in production of each crop can also give a rough for-ward estimate.

The bottom line calculated from the above data, for each material, can be referred to a table of nutrient values to determine the metabolisable energy (ME) and protein value (CP) of each. The sum is the maximum potential feed resource base (FRB), without scavenging feed. Some of the figures may be rubbery; but they will put you in the general area. Things will change; a new crop may come in, so redo the calculation every couple of years.

The Consumption Demand for Feed in a Project

- How many hens, growers, chicks and cocks are expected in the average family flock?
- What are the anticipated average body weights average growth rates and average hen day production percentages (HDP%) of each class of bird?

Apply the formula:

$$\text{ME/bird daily} = W^{0.75}(173\text{-}1.95T) + 5.5\Delta W + 2.07\ EE$$

to each class of birds (National Research Council 1994).

W = body weight (kg)

T = ambient temperature (^{0}C)

ΔW = change in body weight (g/day)

EE = egg mass (g/day)

Multiply the result of the calculation for each class by the number of birds in that class, and add all the totals to give the ME requirement for each family flock per day.

Multiply by 365 to give the family flock requirement per year. Multiply by the number of families to give the project requirement per year. Knowing the energy values of your feed from standard tables, you can calculate the weight of feed required by each family per year.

Divide the supply of feed per year by the requirement of each family per year, and you have the number of families the project can support in a small-scale intensive system.

If the scavenging feed resource base (SFRB) is to be utilised as part of the nutrition for the project, then use the formula above to calculate the existing SFRB for each family, multiply by the number of families which will be in the project and add it to the supply of feed available for the project. This will be a slight overestimate because the feed from the environment will have to go further when the project starts, but that is not a major effect.

If the project is already running, then the demand can still be calculated as above. During the calculation any drain of resources from other regions will become apparent, and should be taken into consideration. The contribution of the SFRB can be assessed by examining the feed in crops at different times of the day, to determine how much is household refuse, how much from the environment and how much is supplementary feed. Those ratios can be used to calculate the actual feed resources available to the project in areas to be added to the project. To add new families in the same area the figure will be lower, because the SFRB will be shared among more families.

The end figure is the demand for feed resources for the project, and the difference between supply and demand indicates how fast the clock is ticking. Is there a source of animal protein in the FRB? A small quantity goes a long way towards improving the efficiency of utilisation of protein in the FRB for growth and egg production, because it provides essential amino acids. There will be seasonal fluctuations in the capacity of the SFRB in line with periods of fallow or flooding, cultivation, harvesting and processing. Allowance and adjustment should be made for these as discussed below.

The size of the SFRB and levels of major nutrients have been measured in a range of environments. The proportion of the SFRB provided by household refuse is directly related to the density of housing and usually household refuse provides most of the SFRB. There are substantial differences between countries, in the amount of scavenging feed available; but on limited data, the nutrient values are similar.

Table: *Analyses of crop content of scavenging village hens at four locations and during wet and dry seasons. The crude protein requirements for an average family flock are included as a measure of biomass*

	Crop content		Scavenging feed resource base			Flock biomass crude protein
	Crude protein %	*ME Kcal/kg*	*Household: Environment %*	*Dry weight (g/day)*	*Crude protein (g/day)*	*(g/day)*
	9.4	-	0.72	534	50	44
	7.6	2130	0.55	260	20	91
	9.3	2220	0.32	834	78	75
	9.8	2430	0.85	639	63	83
	8.1	2480	0.91	603	49	36
	9.7	2190	0.72	575	56	57
	11.8	2230	0.84	365	43	48
Mean	9.4	2280	0.7	544	51	62

Sources: Gunaratne et al., 1993, and Roberts, 1995.

Village hens have similar growth characteristics, and nutrient requirements, to those of commercial white egg layers. The main differences between the two, nutrient requirements for egg production and the number of eggs produced probably come from their histories. Commercial egg layers have been selected to maximise egg production and have been provided with the appropriate nutrients. Whereas village hens have been subject to natural selection to maximise egg production on the diet available to them in the scavenging system.

Nutrient Requirements for Poultry Production

The approximate nutrient requirements of commercial and village chickens are set out in Table below:

Table : *Approximate compositions of diets required by different types of chickens*

Age of birds and supply from the SFRB		**Metab. energy** *(kcal/kg)*	**Crude protein** *(%)*	**Egg production** *(HDP%)*
	White egg layers	2850	16	-
Immature:	Broilers	3200	20-23	-
	Village chickens	2800	16	-
	White egg layers	2900	15	80
Mature:	Broilers	3200	10	30
	Village egg layers	2300	10	30
Available from the SFRB in a village		2300	9.4	28

Table: *Nutrient utilisation efficiency of Sri Lankan village chickens for growth in pens over 17 weeks and for egg production over 40 weeks*

Feeding system	**Growth**				
	Weight gain (g)	Nutrient intake		Nutrient efficiency	
		ME (kcal/bird)	*Crude prot. (g/bird)*	*ME (kcal/g gain)*	*Crude prot. (g/g gain)*
Free choice	1,648	21,729	1,169	13.2	0.71
Commercial	1,498	24,074	1,463	16.1	0.98
Feeding system	**Egg production**				
	Egg weight (g)	Egg production		Nutrient efficiency	
		HDP%	*g/bird/day*	*ME (kcal/g egg)*	*Crude prot. (g/g egg)*
Free choice	48.2	27	13	16	0.78
Commercial	46.5	33	15	19	1.26

Sources: Chandrasiri et al., 1994 and Wickramaratne et al., 1996. Sri Lankan village hens had good reserves of abdominal and subcutaneous fat when scavenging (Gunaratne et al., 1993). Egg production may respond to improved nutrition, in other strains of village chickens (Horst, 1991). Hen day production of 12 to 20% for scavenging village hens in other regions, improved marginally with better nutrition in the village (Avante, 1989), and rose to 40% when Indonesian village birds were in pens with excellent nutrition.

The protein requirements must be qualified by the need for essential amino acids. Monogastric animals have to obtain them from animal sources like fishmeal, meat meal and small metazoans. If the diet is deficient in essential amino acids, then utilisation of protein is inefficient and therefore expensive. Vitamins are usually available in a mixed diet, which includes green feed. If there is more than about 7% of fibre in the diet the efficiency of utilisation falls.

Table: *Feed intake, growth rate and feed conversion ratios of village chickens and hybrid layers under different husbandry conditions*

Measurements	**Village chicken**			**Hybrid layer**
	Husbandry			Free choice in pen
	Scavenger	*In pen*		
		Conventional	*Free choice*	
Age at point of lay (days)	197	133	143	130
Weight at point of lay (g)	1227	1600	1510	1520
Growth rate (g/day)	9.2	12.9	11.8	11.7
Feed intake (g/day)			59.0	60.3
Feed conversion ratio			4.5	4.1
Mortality rate (%)	40	8	8	3

Source: Chandrasiri et al., 1993, and Wickramaratne et al., 1993.

If diets are compounded in the village, it is not realistic to expect precise balance and optimal utilisation; but a reasonable approximation is achievable using tables. In villages, free choice feeding has much to offer, as the birds select their own diet from different feeders, one with high energy materials, and the other a high protein source and minerals. The birds match their intake to requirements. In the study of Chandrasiri et al., (1994), village chickens chose diets for growth and egg laying, which was substantially lower in protein than is provided in commercial diets.

Optimal Utilisation of the SFRB by Smallholders

The SFRB is not the exclusive domain of scavenging chickens. They compete with semidomestic pigs, goats and dogs, and a range of wild scavengers. Nevertheless, when the SFRB is referred to in this paper, that portion utilised by scavenging chickens is implied.

The model is that of a wild population with no territorial or reproductive constraints. The population reproduces until a Malthusian constraint intervenes, and war, disease or starvation then reduces the size of the population until it is within the capacity of the feed resources. The weakest members of the population die, and productivity is compromised. The pressures of competition have diverted productive yield into waste.

Figure: *Model of a wild population with no territorial or reproductive constraints*

SCAVENGING VILLAGE CHICKENS				
Malthusian constraints		Type of flock		Use of products
	Family flock		VILLAGE FLOCK	
	Σsfrb	=	SFRB	
	⇓		⇓	
	Σflock	=	FLOCK	
		⇓		
		PRODUCTION		
		⇓		
	Complete	⇐	Control	
	⇓	⇐ ⇒	⇓	
	WASTAGE	⇔	YIELD	
Starvation	Eggs		Eggs	Consume
Predation	Chicks		Chicks	Sales
Disease	Growers		Growers	Gifts
Mortality	Laying time			
	(broody)			

The most vulnerable members of the population are the chicks and growers, probably because the gap between their requirement for protein (table above) and the availability of nutrients in the SFRB (table below), is greater than for any other class in the population. Moreover, physically they are the weakest members of the population. Low survival rates for chicks and growers have been reported in Indonesia (Kingston, 1980), Thailand (Janviriyasopak et al., 1989) and Sri Lanka (Gunaratne et al., 1993), and were alleviated by supplementing chicks in the field.

Figure: *Survival rate of village chickens Creep feeder arrangement*

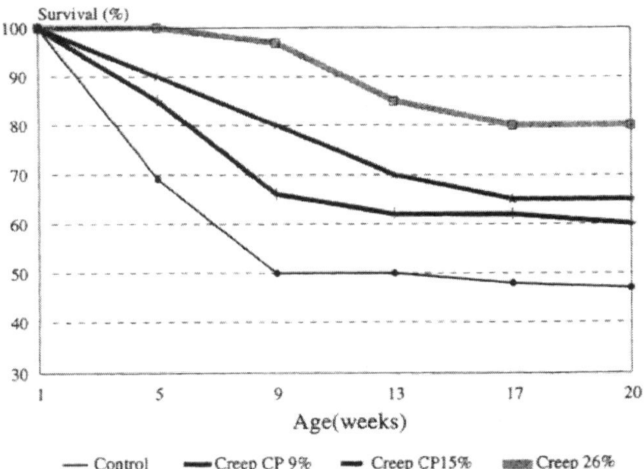

Figure: *Survival of village chicks and growers from hatching to 160 days: Indonesia and Sri Lanka*

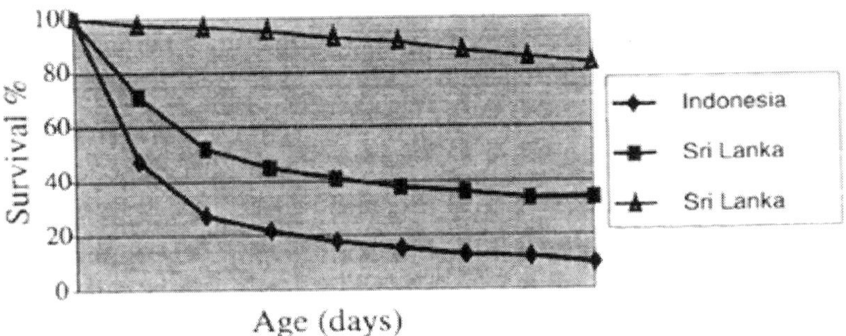

Growth rates of village chicks and growers were much higher in an intensive system than in villages.

Disease is not commonly associated with the high mortality, the steady rate of attrition being unlike that of an infectious or parasitic disease. The primary cause is starvation, the chicks and growers which die tending to have lower growth rates than the average for their brood (Wickramaratne et al, 1993).

The growth rates of survivors to 70 days ranged from 2 to 7 g/day with a mean of 4.4 (Gunaratne et al., 1993). Chicks and growers which die, also tend to be those with lighter feather colours, that is, those which are more conspicuous (Wickramaratne et al. 1993). Predation is the ultimate fate of most chicks and growers.

Presumably anything which weakens a chick, such as inadequate nutrition, infection or parasites, increases the likelihood of it being taken by a predator.

Placing household refuse in a creep feeder for chicks in villages, for a short period, twice a day, increased the survival rate; but did not improve the growth rate. Supplementing the household refuse with protein improved both survival rate and growth rate.

Growth Rate of Village Chickens

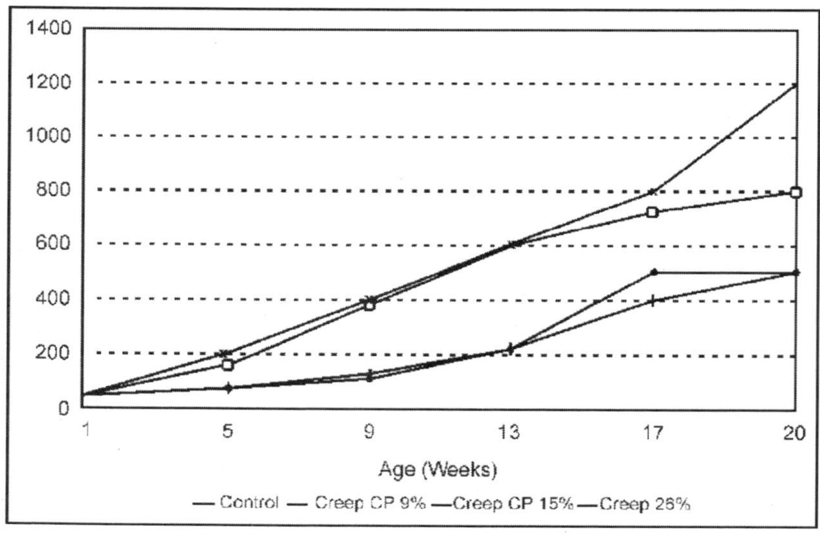

Figure : *Creep feeder arrangement*

There was a very high survival rate and improved growth rate, when village chicks were reared in a pen with free choice feeding (Chandrasiri et al., 1994).

In the creep feeder study, all of the household refuse was supplemented and the substantial residue after the chicks had fed was made available to the remainder of the flock, which showed no increase in productivity. It would be expected that, if a small amount of household refuse were supplemented and continuously available to the chicks, then there would be a greater benefit at lower cost.

It is important to note that if more chicks and growers survive, then the capacity of the SFRB will be exceeded, unless countermeasures are implemented.

If the families consume more eggs (hatch fewer chicks), or consume more growers, they will benefit, and the biomass of the flock will remain in balance with the capacity of the SFRB.

The scavenging model can be used to interpret the differences between the meat production systems described in Indonesia by Kingston & Cresswell (1982) and in Thailand by Janviriyasopak et al., (1989), with the predominantly egg production system favoured in Sri Lanka (Gunaratne et al., 1993). In the meat production systems there are a large number of hens, a high proportion of eggs incubated, and very low survival rates for chicks and growers.

It is suggested that families are seeking to produce more meat by rearing more chicks; but the attempt to raise production increases the competition pressure within the village flock and reduces the survival rates of all chicks and growers. In the egg production system there are fewer hens, higher hen day production and a higher proportion of eggs is consumed. It is suggested that the competition pressure is reduced because the higher egg consumption reduces the number of eggs being incubated and chicks hatched, so a higher proportion of chicks and growers survive.

Table: *Average constitution of family flocks of scavenging chickens in villages of different countries. The value for the family SFRB is calculated from the constitution of the flock and the growth rate of each class of bird as described by Roberts (1992)*

Country	**Number of birds in each class**				**Family SFRB (kg/year)**	**References**
	Chicks	*Growers*	*Hens*	*Cocks*		
Indonesia	6.7	6.3	11.4	3.6	475	Kingston & Cresswell (1982)
Thailand	14.0	8.0	4.5	1.4	390	Janviriyasopak et al., (1989)
Sri Lanka	2.4	3.7	4.0	1.4	200	Gunaratne et al., (1993)

***Table :** Egg production by scavenging hens in villages in different countries, and the disposal of the eggs*

Country	Disposal of eggs					References
	Laid	*Consumed*	*Incubated*	*Hatched*	*Matured*	
	Per hen per year					
Indonesia	72	9	63	52	3	Kingston & Cresswell (1982
Thailand	49	0	49	36	3	Janviriyasop et al. (1989)
Sri Lanka	110	102	8	5	0.5	Gunaratne et al. (1993)
	Per family flock per year					
Indonesia	821	103	718	593	27	Kingston & Cresswell (1982)
Thailand	221	0	221	162	13	Janviriyasop et al. (1989)
Sri Lanka	440	408	32	20	2.5	Gunaratne et al. (1993)

Scale of Production within Households

Scavenging chicken production is seen as free because there are no alternative uses for the inputs. However, as we have seen, there is a low ceiling on the level of production attainable from scavenging. Moreover there is a large step up to small scale intensive production. For a wide variety of reasons many village families don't want to take that step, or cannot take it.

Many of them obtain great satisfaction from the existing scavenging system and want it to stay as it is. Others might be encouraged to take small steps to increase the productivity of the SFRB. Some of the pros and cons of the existing scavenging system are considered below, and some low cost options for addressing the problems are suggested.

Favourable aspects of the existing scavenging system for village families:

- They own their own chickens and can control their reproduction by controlling incubation of eggs. This also means that the cost of re-entry is low if the family flock is wiped out by disease.

- The perceived superiority of village birds and eggs means that consumers will pay a premium over the price of commercial produce. The premium may be two to three fold (Indonesia) or 10 to 20% (Sri Lanka).
- They control the time and place of distribution of their own household refuse.
- They have access to the environmental components of the SFRB.
- They have access to the household refuse of households, which do not keep chick-ens.
- The quality of the SFRB is adequate for egg laying by village hens; but at a low level of production.

Deficiencies of the existing scavenging system for village families:

- The capacity of the SFRB is fixed and low.
- Families have no control over the allocation of the SFRB. If a few families show forbearance in order to relieve the competitive pressures in the system, the slack is rapidly taken up by the flocks of families, which do not show restraint. The only thing, which has changed, is the distribution of the SFRB between the village families.
- Survival rates of chicks and growers are unacceptably low:
  - The quality of the SFRB is inadequate for chicks and growers.
  - Broody hens are inefficient when raising a clutch of chicks.
- There is excessive downtime while hens are broody.
- Eggs are wasted:
  - Hens lay away from the homestead.
  - Hatch rates are low.

Addressing the Problems

- Establish some control over the SFRB by restricting access to the families' house-holds refuse, perhaps by withholding it until the birds return in the evening. It is possible to maintain about three village hens on the refuse from one house-hold. They could be kept in a small pen suspended on the side of the house, similar to that described by Macgregor & Abrams (1992), and given some access to scavenging. The nutrient values of household refuse approximate the optimum for egg production by village hens. If the number of village hens were appropriate for the amount of household refuse, good production, by village standards, would be expected.

- Raise a few chicks and growers with care. Fifty% survival rate to 12 weeks should be easily attainable with preferential access to supplemented household refuse. Use a creep feeder, or confine them in a pen with supplemented house-hold refuse for most of the day. Put the remainder of the flock in the pen each evening. In the average family flock in West Java, more than 700 eggs are incubated to produce 27 mature birds/year. At least 600 extra eggs would be at the disposal of the family if only a few chicks were reared with care. In addition the proportion of SFRB wasted on dead chicks and growers could be converted to yield more growers or more laying hens.
- Rear chicks and growers during seasonal highs in the capacity of the SFRB, such as during times of cultivation, harvest or processing, depending on the local situation.
- Restrict the broody time of the hen. The shorter the broody time the higher the egg production (Prasetyo et al., 1985). Each household would have a different stage at which the benefit of having the hen with the eggs or chicks was outweighed by loss of egg production. Hens lay for longer if eggs are removed daily. Crossbred and hybrid hens have much shorter broody periods and higher egg production; but require a correspondingly higher quality diet.

Table : *The effect of broodiness on egg production of village hens*

Hen activities			***Production parametres for one year***	
Lay	***Hatch***	***Rear***	***Broodiness in days***	***Egg production in numbers***
+	+	+	257	52
+	+	-	145	115
+	-	-	125	132

Source: Prasetyo et al., 1985.

- Have adequate nesting space and train pullets to use it. It is not difficult.
- Treat eggs for incubation with care and store them for a minimum period. In Tanzania the hatchability of freshly collected eggs was 93% and decreased by 5.3% for each day of storage. The body weights of chicks hatched from stored eggs were also lower (Abdou et al., 1990). Thus if eggs are stored for a week before being placed under a broody hen, about 30% are wasted, and the hatched chicks are probably weaker.
- Cull unproductive hens

- Provide information. There are many simple devices and procedures for egg laying, hatching, rearing, feeding and protecting chickens in villages; but many are known only in local communities. Well illustrated pamphlets, audiovisuals, demonstration plots and key farmers can all make valuable contributions, and benefit many families in other regions and countries. Information concerning chicken nutrition, breeding, health, husbandry and production can also influence contraception, conception, childbearing, lactation, growth and development of children. Chicken production being the source of high quality protein needed by the family.

Considerable effort has gone into attempts to upgrade the scavenging system; but the medium term failure rate of apparently successful poultry projects in villages is high. We have never had funds for a development project, and so have not promoted any of the procedures described above.

However, we have done research on some of them and have undertaken extensive village trials on hand rearing, creep feeding and supplementation. Village families have usually cooperated willingly and been interested in the results. Use of a procedure has always ceased shortly after a trial. Motivation of the village family may be the main constraint to increasing the productivity of the basic scavenging system. Village families are getting something for nothing now with little risk - if the system collapses then it can be started again at little cost. Cock fighting is still alive and well in many villages.

Many will be unwilling, or unable, to undertake the effort and risk associated with inputs to the scavenging system. Productivity is irrelevant for those for whom the social and cultural contributions of the traditional village chickens are their most important attributes. They believe, with some justification, that scavenging chickens are the ultimate poultry.

Small Scale Intensive Poultry Production in Villages

For those village families, which want to go to a level of production, which is higher than that, available from scavenging, more, and higher quality feed must be provided. Once one significant input is provided, a cascade of further inputs becomes mandatory. There is no point in providing better feed to village hens which cannot utilise it efficiently, so specialised birds must be obtained. With more valuable birds it is necessary to keep them secure, and to protect them against disease, particularly Newcastle disease. If many families increase production then new markets may be necessary so storage and transport become issues. This scenario is highly commendable, provided it has been thought through. Are all of the inputs available? And is there an infrastructure to maintain the supply?. Establishment of an intensive husbandry system could be facilitated by an appropriate production package (Johnston & Cumming, 1992). A small-scale intensive industry is more likely to succeed if there is a reliable local supply of a high energy feed material such as rice bran, coconut meal or a seed meal.

Preston, (1992) has proposed intensive poultry production based on oil palm oil or seed. If sufficient households in a community opt for small scale intensive production there is then scope for other families to contribute and benefit by providing supporting activities such as breeding and rearing of replacement chicks, vaccine storage, vaccination, collection and distribution of feed components, supply of materials, and product marketing.

There are several possible bases on which sustainable small scale intensive chicken production can be established in villages. Governments can do it through banks and livestock services; but the resulting projects are rarely sustainable (Haan, 1992). It is also possible for village production units to be integrated into a vertical industrial system including feed mill, hatchery, hybrid birds, veterinary services, abattoir and marketing, with the parent organisation providing some credit and taking some risk, and profit (Bhannasiri, 1992).

In effect the village family provides housing and labour; but in the process is vulnerable to the vagaries of the commercial industry and the market. In Turkey similar vertical integration has been achieved through cooperatives which have developed in place of the commercial intensive industry (Unver, 1992). Project aid could be a good option if an adequate infrastructure is established before the assistance is withdrawn. A poultry cooperative could underpin the operation, and poultry shops can also provide the support system with feed, credit, vaccines and marketing (Sonaiya, 1992).

Pens are necessary to protect the birds from predation and theft. The main requirements are for protection from direct sun, adequate space, ventilation, perches, feeders, nests and water. An elevated slat floor (Owoade & Oduye, 1992) can assist ventilation and disease control, and make the faeces available as fertiliser, or for fish culture. Split bamboo is an effective building material.

It is usually better to have an adequate hen house and no yard, than to have a small hen house with a fenced yard, which rapidly becomes a desert. Fencing may be more efficiently used to protect a kitchen garden fertilised with the manure.

Effort should be made to ensure that the important role of women in village chicken husbandry is expanded as small-scale intensive village production develops.

Genotype

The genotype of the chickens is an important component in efforts to optimise utilisation of the SFRB, regardless of the intensity of production. Market forces should determine the type of bird used. Where the local cooking style includes chicken boiled for long periods, flavour is important and if consumers have the money, there is commonly a high premium on the price of traditional village chick-ens. In such circumstances it may be worthwhile to accept, or adapt to, the disadvantages of village chickens in the intensive situation, particularly the stress and fighting problems caused by crowding (Lee, 1992), and the inefficient utilisation of nutrient materials for egg and meat production.

Village chickens, and their crosses with improved breeds, have been selected for productivity characteristics on high quality diets. Simulated village chickens have been developed in Taiwan by crossing village and commercial birds and selecting the offspring for the overt features of the village birds such as coloured plumage, bare necks, pigmented bones and skin, pigmented and variant combs and long shanks, together with the efficient production characteristics of the commercial birds (Lee, 1989).

Production is medium to large scale intensive on high quality diets, and the product is well received, so the enterprises are highly profitable. While there may be significant heterosis benefits in growth rates and egg production for village X layer hens (Horst, 1988, 1991); it would be difficult to maintain heterosis benefits in a village breeding system.

Where there is only a small premium for traditional birds then commercial crossbred layers, broilers, or a suitable dual-purpose breed are more efficient, both nutritionally and economically, for small-scale intensive village production. Production from such birds has been successful when introduced to the village as day old chicks (Roberts & Senaratne, 1992) or as young adults (Huchzermeyer, 1973). With supplemented scavenging, hen day production averaged up to 60% with eggs weighing about 60g. However there was a high attrition rate attributed to theft and predation.

The main disadvantage is supply of stock, because commercial interests control the genetic material for producing hybrids, and they are geared to supplying the large-scale industry.

They tend to be apathetic, or even antagonistic, to orders for fewer than thousands of chicks. The dual purpose breeds are no longer fashion-able in developed countries; but could be appropriate in villages because village people desire both egg and meat production, and could breed their own replace-ment stock.

Where scavenging is to be incorporated, a cross with a dual-purpose breed could be more appropriate.

In the past there have been attempts to improve the productivity of scavenging village chickens by 'genetic upgrading' with schemes like cockerel exchanges.

It became apparent that birds, which were highly productive in an intensive environment with high quality nutrition, did not survive in the rough and tumble of the village with its low quality SFRB, and aggressive cocks.

Protein is the most expensive component of the diet. When two lines of chick-ens were selected for growth rates on balanced and on low protein diets, growth rates of the lines were similar when both were on balanced diets; but chickens of the line selected on the low protein diet were 10 to 20% heavier at six weeks on the low protein diet (Sørensen, 1985).

The adaptation of village layers to the low protein scavenging diet, is probably a similar phenomenon. In many regions there could be a case for crossing village chickens with a coloured dual purpose breed, and selecting progeny for locally desirable morphological characteristics, and for productivity on an *economical, locally available* diet, or more simply on a low protein diet.

The strain could be valuable for controlled scavenging or small-scale intensive chicken production.

The multiplication and production systems used by Fattah and Swan (1999), to produce chickens for the Participatory Livestock Development Project in Bangladesh could be used to test genotypes for suitability for a supplemented scavenging diet. Eggs are produced in villages, or on government farms, hatched and reared in villages, and sold to village producers.

Birds being assessed could be on appropriate diets and production would be measured at each stage. When the best genotype is selected, it could be bred through the F2 generation if a crossbred. Chickens could then be selected for productivity on lower quality feed.

Producers may need a small subsidy; but the overriding advantage would be selection in the environment in which the chickens are to produce. There is little point in selecting chickens on an institutional farm if they are to scavenge in villages. Higher productivity on a low quality diet would provide a substantial benefit over the large number of birds in a project.

There is flexibility in the scale of operation of each village family, the optimum combination of the factors of production being relatively independent of scale when production is intensive. At the lower end two to four commercial layers in a small unit suspended from the wall of the home, could be fed on the household refuse appropriately supplemented.

Flocks around the size of present village family flocks of about four hens, a cock and some chicks and growers can be housed and fed household refuses and supplement; but still allowed scavenging time.

In larger flocks household refuses ceases to make a significant contribution to the diet, so the family has to rely on the by product and feed markets. When the level of production in the community exceeds the local demand, then the producers are competing with large commercial producers for market share.

The advantages, which the small-scale intensive producer in the village could have, are low expectations of return for labour, small capital investment, and low costs for purchase, transport and processing of feed, if crop by products are available locally.

They also have cheap access to local markets for the sale of produce, for example the village farmer can sell live chickens locally for meat, whereas the large scale producer has to have access to an abattoir, a freezer and a refrigerated or insulated distribution system.

If we accept that the objectives to be promoted by village chicken production are family nutrition, family opportunities and satisfaction of social and cultural aspirations, then traditional scavenging and small scale intensive production have most to offer.

For the foreseeable future, in the villages of the developing world, there will be many families still keeping scavenging chickens, some because they like having them around, and some because they need their products.

Table : *Characteristics of different levels of chicken production for village families in developing countries*

Characteristic	Traditional scavenging *(1 to 10 hens)*	Small scale intensive *(10 to 50 hens)*	Medium scale intensive *(50 to 1000 hens)*
Households that benefit from production	Large number of poor rural families	Moderate number of rural families with education or initiative	A few educated families such as public servants, businessmen and professionals
Benefits	Managed by women. Family nutrition, small cash flow, meet social obligations and satisfy cultural needs.	Women's opportunities. Nutrition, income and family nutrition.	Income
Production consumed by	Poor rural families	Local families	Local urban dwellers
Resources utilized	Material from the environment and refuse from the household	Byproducts of local crops and industries	Byproducts of local and regional crops and industries. Commercial feed
Marketing requirements	Personal contact and middleman	Middlemen and local retail outlets	Middlemen and regional retail outlets
Infrastructure required	None. Information could help	Sources of information, birds, feed, vaccines, medicines. Markets. Possibly credit	Sources of information, birds, feed, vaccines, medicines. Markets. Credit.

Other Poultry Species used for Scavenging Production in the Villages of the Developing World

Muscovy ducks (*Cairina muschata*) and mallard ducks (*Anas platyrhynchos; A. pocilorhyncha*) can utilise small fish, snails and water weeds from paddy fields, irrigation channels and swamps, so their SFRB is largely separate from that of chickens. There is a sophisticated mallard duck husbandry system in association with wetland rice cultivation in some regions of South and Southeast Asia. Ducks scavenge newly cultivated and newly harvested fields; but unlike chickens the scavenging is supervised. Eggs are incubated in traditional solar heated sand incubators, and ducklings are sexed shortly after hatching - by the note of their 'cheep'. The newly hatched ducklings are imprinted to the individual totem of each owner, and it is used to control them all the days of their lives.

On Bali, flocks are kept at one location; but on Java, husbandmen may migrate hundreds of kilometres following cultivation and harvest and selling the eggs. At the end of the journey the ducks are sold for meat, and the herdsman recommences with a young flock. Ducks also integrate well with fish culture. They are good egg layers, and all eggs are laid shortly after dawn, before the birds are released from their portable pens. There are many suitable environments in other regions; but ducks are not kept, possibly because of the need to transfer a production package, and because local palates do not appreciate duck eggs and meat.

Geese (*Anser cygnoides*), Guinea fowl (*Numida melagris; N. ptilorhynca*), Turkeys (*Meliagris gallpavo*), Pigeons (*Columba livia*) and Japanese Quail (*Coturnix coturnix japonica*) all have regional popularity; but not wide appeal.

The Hosts

It is probable that most avian hosts, domestic and feral, can be infected with strains of Newcastle disease virus. Kaleta and Baldauf (1988) listed more than 250 species of free-living and caged birds that have been infected with Newcastle disease virus. This is in addition to the common species of domestic birds that become infected. The consequences of these infections will vary with the strain of virus and the species of host. Chickens are the most important host for Newcastle disease virus. However, village flocks are not limited to domestic chickens. We must be aware of the possible role of other avian species.

Ducks deserve special consideration. Ducks are reported to be readily infected with Newcastle disease virus and to be capable of spreading the virus. There are few reports of clinical Newcastle disease virus in ducks, but we must be wary.

There are several accounts indicating a high rate of carriage of Newcastle disease virus by ducks. For example, in a village situation in Indonesia Kingston and Dharsana (1979) found that the virus persisted for one year in a flock of only 300 ducks. In work with which the author has been associated in Vietnam, strains of Newcastle disease virus that are virulent for chickens have been isolated from ducks.

Some isolates of duck plague virus (a herpesvirus) have been contaminated with Newcastle disease virus. It would be useful to determine whether the two viruses act synergistically in dually infected ducks. These findings were with Pekin ducks. Even less is known of Newcastle disease in the Muscovy ducks that seem to be more common in Africa. At a recent IAEA/FAO meeting in Morocco, it was observed that in Tanzania Newcastle disease is a greater problem in chickens in villages where ducks are also kept.

Turkeys can also be infected with Newcastle disease virus, but are apparently not very susceptible to disease. Certain strains of Newcastle disease virus have become adapted to pigeons in some countries and cause disease in both pigeons and chickens. Accounts of the deaths of pigeons in African villages during out-breaks of Newcastle disease are not unknown. Veterinary authorities in Africa give very diverse views on Newcastle disease in guinea fowl. Opinions vary, from these birds being entirely refractory to guinea fowl being highly susceptible and a danger to commercial chickens. From China there has been a report of a strain of Newcastle disease virus adapted to, and causing disease in, geese (Linchuan and Hong, 1998).

Newcastle disease virus can infect mammals. Human infection occurs and, at least with virulent strains of the virus, causes severe conjunctivitis. There has been an isolation of Newcastle disease virus from a pig in Indonesia.

There may be more unusual hosts for Newcastle disease virus. Again from Indonesia there has been an account of the apparent replication of Newcastle disease virus in rice field crabs (Kingston and Dharsana, 1977).

The Virus-host Interaction

Newcastle disease virus reacts with avian hosts in various ways. When non-immune domestic chickens and highly pathogenic strains of Newcastle disease virus meet, the common sequel is an acute disease with mortality close to 100%. Surviving birds will often have defects that are attributed to neural lesions. The birds will often be paralysed or have twisted necks. This severe disease, attributed to strains of Newcastle disease virus that we call *velogenic*, occurs in both village and commercial poultry. It is the only form of the disease that is of consequence in village chickens. The syndromes attributable to viral strains of lesser virulence would not be noticed in village flocks.

In commercial chickens, strains of Newcastle disease virus of moderate virulence (*mesogenic* strains) cause lower rates of mortality in mature chickens, but severely deplete egg production. Strains of low virulence (*lentogenic* strains) cause little mortality except in young birds, but do reduce egg production. These strains also interact with other pathogens, especially respiratory pathogens, to produce severe clinical disease.

Most Newcastle disease vaccines are living vaccines, drawn from the ranks of lentogenic or even mesogenic viruses. Their use may alter the epidemiology of Newcastle disease. We now also recognise avirulent strains of Newcastle disease virus that cause no clinical signs when they spread between chickens of any age by natural routes. Lentogenic, mesogenic and avirulent strains also circulate naturally in some countries, modifying the epidemiology of Newcastle disease.

Most strains of Newcastle disease virus spread by the respiratory route, infected chickens producing aerosols of infectious virus. Faecal excretion also occurs. With some strains this is probably the most important route of spread, with infection being by ingestion.

There is little evidence for the spread of Newcastle disease virus through the egg (true vertical transmission), though transmission of virus on the shell of infected eggs is well recognised. Embryos inoculated at about 9 days of incubation with avirulent viruses (that do not kill embryos by this route) have been allowed to hatch. The chicks are infected (French *et al.*, 1967). However the same viruses given by yolk sac route at 4 days of incubation kill the embryos (Kim and Spradbrow, 1978). The real experiment, with collection of eggs from viraemic hens, seems not to have been done.

Human intervention is probably a common source of infection. The clothing and shoes of people working with infected poultry are readily contaminated. Food, water and items of equipment used on infectious premises are also sources of mechanical spread of contagious virus.

Infected Chickens are the Usual Source of Infection in Villages

I believe that most of the cases of Newcastle disease that we see in village chickens can be attributed to chickens that are shedding virus. These will usually be chickens that are incubating the disease. Virus excretion commences before clinical signs occur. They may however be birds that have recovered from clinical infection, or vaccinated birds. Vaccinated birds may show no clinical signs at all on challenge with virulent virus, but they will become infected and excrete the virulent virus. They can transmit the virus to other birds by contact for up to two weeks.

In many countries, seasonal outbreaks of Newcastle disease are recognised. Attempts are often made to attribute these outbreaks to the weather conditions prevalent at the time. There may, however, be more realistic explanations based on patterns of movement in chickens and changes in the volume of chicken markets.

Accordingly, the seasonal outbreaks of Newcastle disease might not be directly due to prevailing climatic conditions. Possibly the real correlation is with the planting season and the need for villagers to sell chickens to buy seed rice. The volume of chickens traded through the markets increases at this time. From Uganda it has been suggested that outbreaks that occur in the dry season are not really because the virus survives better under these conditions. This is the time of low employment in the agricultural sector. Villagers use the spare time to visit kinfolk, and usually carry chickens as gifts. Outbreaks in other countries might be related to marketing for festivals, rather than to the season. There are, for example, the out-breaks that occur before Christmas in Ghana and before Easter in Ethiopia.

In many countries veterinary workers, and indeed village people, recognise that the introduction of new birds to a flock is often associated with an outbreak of Newcastle disease. This lesson needs to be spread by extension workers, although it is not possible to restrict the entry of new chickens into village flocks.

Villagers become astute at recognising the early signs of Newcastle disease. There are two common responses. One is to salvage sick or even dead birds through the cooking pot. We cannot disapprove of this practice when so many villagers are always hungry. We can however suggest that the infected entrails are buried, and not discarded to be scavenged by other creatures.

The second remedy is the market place. Markets must be a common source of Newcastle disease infection, sometimes through the random sale of infected birds, but more commonly through salvage sales. Huchzermeyer (1993) made the interesting suggestion that Newcastle disease virus may spread amongst village chickens at night rather than during the day when ultraviolet radiation is strong. He believed that chickens that were housed at night would be more readily infected than birds that roosted in trees.

With roosting flocks, the inability of sick birds to fly to branches would be an added protection. Brooding hens and hens with chickens would also segregate and protect them-selves. Huchzermeyer also postulated a state of endemic Newcastle disease that would not depend on persistent carriers.

Epidemic Newcastle Disease in Village Flocks

In many areas, the chickens in small village flocks have no previous experience of Newcastle disease virus. They lack antibody and when the virus is introduced, it spreads quickly. Then we see the epidemic occurrence of Newcastle disease. These are the flocks where we will see adult birds dying as we walk through the village. These are the events that colour our thinking about Newcastle disease, and the attitude of the villagers to the disease. There will be few survivors. These are the events that ensure that in all languages and dialects there is a word that translates into English as Newcastle disease.

Epidemic Newcastle disease is not self-sustaining. The virus will vanish when no susceptible chickens remain, or when the chicken population contains too few susceptibles for the virus to spread. We must envisage other situations in which populations of chickens maintain endemic Newcastle disease, or in which other hosts maintain the virus.

Endemic Newcastle Disease in Village Flocks

When Johnston (1990) commenced to monitor village flocks very closely for New-castle disease, he found that epidemic spread was not the only form of the disease. He noted that Newcastle disease "may be able to smoulder enzootically for several months, even years, in a typical village housing one or two thousand birds". The losses of small numbers of chickens, against the general background mortality, will not attract attention. The author noted that monitoring processes designed to study fast-spreading epidemic disease would not be sufficient to detect endemic infection.

Computer models generated by Johnston (1990) accorded with his field observations. The model indicated that velogenic virus would remain endemic in a group of 5 flocks, each of 80 chickens, sharing the same scavenging environment. Flocks with endemic Newcastle disease will have a mix of chickens with antibodies and chickens lacking antibodies. The problem birds are those that are clinically healthy but incubating the virus and either shedding virus or soon to shed virus. Vaccinated birds can also become infected and shed virus although they develop no clinical signs.

There are practical consequences. Village vaccine trials often involve buyback challenge; in which vaccinated birds and control birds are purchased and returned to the laboratory for artificial challenge. Sometimes one of the apparently healthy birds will initiate the challenge before the intended challenge virus is introduced. This has occurred in the Philippines (Fontanilla *et al.*, 1994). Unvaccinated village chickens are sometimes purchased and housed together for later vaccine experiments. One chicken excreting Newcastle disease virus can kill the whole group. This has happened in Ethiopia. It is better to obtain potential experimental chickens as one day olds and to rear them in isolation.

In Tanzania the problem has been met by holding the purchased chickens together for a fortnight, and then selecting those that lack antibodies against Newcastle disease virus. Agencies supporting women's groups will sometimes purchase village chickens and vaccinate them against New-castle disease.

On one occasion, when vaccination and distribution occurred on the same day, the subsequent deaths from Newcastle disease were a tragedy to the new owners. A better approach, as practised by Dr. Robyn Alders in Mozambique, is to vaccinate and hold for a fortnight before distribution to the new owners.

Economics of Newcastle Disease Control in Village Poultry

Until very recently, there were no methods of controlling Newcastle disease in village chickens. The conventional Newcastle disease vaccines that were effective in commercial poultry found little use in villages. The flocks were small, scattered, multi-aged and under minimal control. The vaccines were heat-labile, relatively expensive and produced in large-dose units suitable for large commercial flocks. Their application required physical control over the chickens.

The development of heat-stable vaccines that spread between chickens, such as strains V4 and I_2, have changed this picture. These vaccines can be given on certain foodstuffs if necessary, allowing vaccination of near feral flocks. These vaccines can produce substantial immunity under village conditions. This has been demonstrated in many countries. We have reached a stage where we can recommend and plan for local production and broad distribution of these vaccines.

At this stage of the project, before extensive vaccination is undertaken, we must face questions of cost benefit. There is not a lot of sound information on the productivity and the commercial value of village flocks. We seem to be in urgent need of the assistance of economists in this phase of the project.

We all know in qualitative terms the value of village chickens. Chickens, their meat and their eggs have a potential market value. The flock serves as a reserve of petty cash that can be drawn on for barter or for sale when essential items are required. Chickens serve social and ritual functions that are difficult to value. Chickens also perform waste disposal functions and provide organic fertilizer. However when we approach economists or funding bodies there is a requirement for cost benefit analysis. Few have attempted to derive monetary values for village flocks.

Johnston (1990) analysed data from 697 monthly flock monitoring visits to vilage flocks in Selangor, Malaysia. Outlays on the flocks were M$11.37 per month, the production M$20.46 per month, with an average monthly profit per flock of M$9.09. He cites a similar study from Sri Lanka, where the flocks are devoted to egg production rather than meat production (Veterinary Research Institute, Sri Lanka, 1990). The target population was 156 flocks, whose egg production allowed each person in each household to consume 2.5 eggs/week, which is three times the national average.

A study in Thailand demonstrated the essential wastefulness of village poultry raising. Age specific mortality rates were 25% for chickens less than 2 months old, 7% for the 2 to 6 month age group and 3% for older birds. The average time in a flock was 5.8 months. Production comprised 24.7% of all removals, but 82.2% of removals at greater than 2 months. None of the removals under 2 months of age were for productive purposes and these outnumbered the total removals of older birds. Newcastle disease was a frequent cause of death. Control of Newcastle disease and increased survival of chicks would greatly improve productivity.

Another approach used by Martindah (1991) was to measure village offtake based on hen days, to determine the productivity of adult females. In this study every 100 hens produced for sale 402 growers and 387 adults (presumably per year).

The lesson seems to be that it is difficult to measure the productivity of village flocks, and extremely difficult to place a monetary value on this production. Yet it is necessary to value the enterprise if the response to vaccination is to be measured. Simple census data suffices for villagers and for simple virologists. Villagers will see every chicken saved as a profit, and will probably value the vaccine in chicken equivalents. I believe that in Bangladesh, villagers are prepared to pay a few percent of the potential final price for the chicken to have it vaccinated. Where vaccination programs are introduced it will be necessary to devise methods of cost recovery if production and delivery of vaccine is to be sustainable.

Granting bodies are more likely to demand a commercial evaluation of village vaccine projects. This is not to be encouraged for these data are difficult to generate. Few are known to me. James (1991) described a theoretical framework for achieving these ends.

The project requires vaccinated and control villages and observation for one year or for a number of whole years. Inputs and outputs would need to be recorded at intervals of one or two weeks. Any change in flock size would be recorded, and increase in flock size would be regarded as reinvestment.

Johnston (1990) carried out an examination of the probable economic impact of oral vaccination on whole countries, using both computer modelling techniques and field data. He estimated that the early ACIAR projects in five countries would generate a flow of net benefits of Australian $389 million.

It should be possible to calculate the cost of thermostable Newcastle disease vaccine made under basic conditions in provincial laboratories. A single egg should yield 5 ml of infected allantoic fluid, the equivalent of 5,000 doses of vaccine. There should be minimal costs in diluting and storing this vaccine, which should have a shelf life of about two weeks.

Investigations on Disease Status of Scavenging Poultry in Morogoro, Tanzania and the Significance of Detailed Characterisation of Pathogens Obtained

Village scavenging poultry is the dominant form of poultry keeping in Tanzania and other developing countries. According to National Census of Agriculture (MOA, 1995) out of 27 million poultry in Tanzania, 93% are indigenous chickens. Evidently most poultry products consumed in the country are from an indigenous source, and poultry keeping represent an important source of income to women in villages (Aboul-Ella, 1992). However, despite the economical, social, and other benefits, village poultry has only been accorded limited importance in both disease control and husbandry.

Diseases and poor management are the major constraints to the health and productivity of village poultry. Newcastle disease has been described as a leading killer of village chickens in Tanzania (Yongolo, 1996). A review of village poultry diseases by Gueye (1998) showed that periodic outbreaks of Newcastle disease devastate scavenging chicken populations wherever they are found in the world. A high mortality rate caused by Newcastle disease is thought to take away the focus from other diseases, and thus mask other diseases in village chickens (Bell, 1992). Reduction of mortalities due to Newcastle disease has been reported in vaccinated chickens (Bell, 1992), however; only limited efforts have been done to control this disease in Tanzania.

Other conditions that affect village poultry include Gumboro disease, fowl typhoid, colibacillosis, and infectious coryza (Thitisak et al., 1989; Kelly et al., 1994). However, the reports show that these conditions are occasional causes of mortality, while some of the diseases such as fowl typhoid and mycoplasmosis have been detected by serological methods only. To add information on clinical and pathological importance of different diseases, we designed our study to include postmortems of dead poultry and bacteriological culture.

Traditional management has been found to encourage parasite infestations in village poultry. A study by Msanga and Tungaraza (1985) showed a high incidence of both internal and external parasites in indigenous poultry in Tanzania. Although parasites are rarely the sole cause of mortality, they have been found responsible for weight loss and poor nutritional status of village chickens. High mortalities that might be expected associated with coccidiosis under scavenging conditions have for unknown reasons not yet been reported to the knowledge of the authors.

In the scavenging poultry system, shelter is provided during the nights only, while during the day the flock is let out to search for feed. This leaves unlimited contact between neighbour flocks, also with other animals such as dogs, cats, and pigs kept around. Consequently, transmission of diseases between flocks, attacks and predation of poultry by cats, dogs, and hawks are common (Mwalusanya, 1998). Indications exist that the cats, being natural carriers of *P. multocida,* can be transient carriers of pathogenic strains of *Pasteurella multocida* to poultry (Van Sambeek et al., 1995).

Under village conditions it seems possible for cats, dogs or pigs to transmit *P. multocida* kept in the same environment. Information about exchange of *P. multocida* clones between different domestic animals is rather limited. Therefore, the current study was designed to isolate and compare strains of *P. multocida* from chickens, ducks, dogs and cats. To investigate the effect of environment on the presence of *P. multocida* (Simmensen et al, 1980) the study was done in three climatic zones (hot, warm and cool).

The objectives of this project were to study the causes of mortality in village poultry by postmortem and to investigate the significance of animals kept in contact in the transmission of *P. multocida* to poultry. Since the last part of these investigations will be published elsewhere (Muhairwa et al., submitted for publication) the present publication will focus on post mortem findings and on the significance of detailed characterisation of pathogens obtained.

The total findings are expected to add to baseline data on diseases of local chickens. These will help formulating control measures for improving the health and productivity of village scavenging chickens with the ultimate aim of poverty alleviation in the rural population.

Materials and Methods

Study area: The study was conducted in Morogoro Region, Tanzania. A study of causes of mortality in chickens was conducted on one village. The sampling of carriers of *P. multocida* was divided into hot, warm and cool zones. In each zone two villages were selected as follows; Kipera and Kongavikenge from hot area (Mlali), Kiroka and Mkuyuni from warm zone (Mkuyuni) and Langali and Nyandira from cool highland area (Mgeta).

Table: *Differences in temperatures, altitude and animal population in three zones*

Zone	*Location*	*Temperature*	*Altitude*	*Chickens*	*Ducks*	*Dogs*	*Cats*	*Pigs*
Hot	Mlali	23-32°C	760m	6674	1800	100	25	-
Warm	Mkuyuni	18-25°C	1200m	9500	1300	88	8	-
Cool	Mgeta	12-18°C	1500m	3440	1200	65	14	200

Post-mortem Examinations

Dead chickens were obtained from one village (Kongavikenge) through visits and assistance of local veterinary extension officer. Post-mortem examinations were conducted as described by Fowler (1996) and appropriate samples taken for microbiological and parasitological evaluation. The data on age of chickens, mortality rate, and size of flock were recorded.

Characterisation of Pathogens and Verification of Diagnosis

Preliminary identification of pure culture isolated from dead chickens was characterised according to methods described by Barrow and Feltham (1993). Subsequent characterisation of isolates suspected to be *E. coli,* were confirmed by reactions of lactose, indole test, Simon citrate, and growth at 44° C in Mackonkey lactose bile broth. Identification of parasites was done by methods described by Soulsby (1982).

Investigation of Carriers of P. Multocida

Sample size, collection of samples, bacterial culture and mouse inoculation were carried out as described by Muhairwa et al.

Identification of Pasteurella Species

Routine identification of *Pasteurella* species included Gram staining, motility test, glucose fermentation test, oxidase and catalase reactions.

The strains obtained were typed by using the following reactions; ornithine decarboxylation, urease production, indole formation and production of acid from sucrose, maltose, mannitol, dulcitol, and sorbitol. Identified species were subsequently brought to Denmark for verification and extended phenotypic characterisation according to Bisgaard et al. (1991)

Statistical Analysis

Statistical analysis of data to determine significance of results was performed using the chi square test.

Results

Post-mortem Findings and Diagnosis

Age of the chickens ranged from one week to approximately 18 months old, adult chickens. Post-mortem signs consistent with Newcastle disease were seen in seven birds, which were obtained from different flocks.

All chickens had hemorrhages in proventriculus just as four chickens had a concurrent *Ascaridia galli* infestation. Increase in the mortality in the flocks of origin of affected chickens was reported. Polyserositis was diagnosed in three chicks from one clutch.

All three carcasses had perihepatitis and air sacculitis. Pure culture of *Escherichia coli* was isolated from livers of these chicks. One case of egg peritonitis was diagnosed in a hen. Mortalities attributable to sticktight fleas (*Echidnophaga gallinacea*) were recorded in five chicks.

These chicks had massive flea infestation on the head, especially around the eye, and pale mucous membranes suggestive of anaemia. Five dead chickens had external and internal injuries including bruises, cuts, liver rupture and fractures. Three of these chickens were also infested with *A. galli* and *Tetrameres americana*.

The results of thirty dead chickens submitted for post-mortem investigations are summarised in Table.

Table : *Results of thirty dead chickens submitted for post-mortem investigations*

	Post-mortem findings/Clinical observations	**Diagnosis**	**Age of birds**	**Cases**
1	Hemorrhages on the proventriculus mucosa, enteritis. Increased mortality and nervous symptoms in the flock. A. galli was found in 4 chickens	Newcastle disease	2-4 months	7
2	Fibrinous airsacculitis fibrinous perihepatitis, hepato-splenomegaly	Colisepticaemia	5-6 weeks	3[1]
3	Peritonitis, inspissated yolk in the abdomen.	Egg peritonitis	18 months	1
4	E. gallinacea. Pale mucous membranes and carcasses.	Ectoparasites	2-4 weeks	5[2]
5	Superficial wounds, fractured bones, ruptured livers, blood in the peritoneum. A. galli and T. americana in 3 chicks.	Fracture of wings, metatarsals and internal injuries	4 weeks-10 months	5
6	Renomegaly, nephropathia	Nephropathia	2-6 months	4
7	Panophthalmitis, rhinitis and conjunctivitis	Infectious coryza ?	4 months	2
8	Loss of feathers, skin granuloma. A. galli found in 1 chicken. Both A. galli and T. americana in 2 chickens	Skin granuloma	3-4 months	3

1. All chicks came from the same clutch.
2. All chicks from the same flock.

Nine cases were not definitely diagnosed. Of these, four chickens had pale kidneys characterised by distinct lobulations (nephropathia). Three chickens had upper respiratory tract infection, with yellowish white flakes in the eyelids.

Bacteriological investigations showed a mixed flora, and further characterisation could not be performed.

The three remaining chicks had granulomatous skin lesions, accompanied by loss of feathers and thickening of the skin, which extended from the head down to the neck.

Two chickens with skin disease were also mixed infested with *A. galli* and *T. americana,* whereas one was infested with *A. galli* only.

Prevalence of P. Multocida in Chickens Ducks Dogs, Cats and Pigs

P. multocida was confirmed in 0.7% of chickens, 7% of ducks 68% of cats and, dogs in all three zones investigated (Muhairwa et al., 1999 submitted for publication).

Prevalence in chickens was significantly lower ($P<0.05$) than all other hosts, whereas prevalence in cats was significantly higher than ($P<0.001$) other animals investigated.

None of the pigs were found carrying *P. multocida*. Other species demonstrated in dogs and cats only included *P. canis, P. stomatis, P. dagmatis,* and an organism of uncertain affiliation tentatively named *Pasturella* taxon 16 (Bisgaard and Mutters 1986).

Occurrence of P. Multocida spp. Multocida in the Three Zones

The prevalence of *P. multocida* ssp. *multocida* in chickens of the warm zone was not statistically different ($P>0.05$) from other zones.

The prevalence of *P. multocida* ssp. *multocida* in ducks of the warm zone was significantly higher ($P<0.001$) than that of ducks in the cool and warm zones.

P. multocida ssp. *multocida* was shown in a single dog each in the hot zone and warm zone, which did not differ significantly ($P>0.05$) from the cool zone.

A carrier rate of *P. multocida* ssp. *multocida* in cool zone cats was significantly higher compared with that of warm and hot zones (Muhairwa et al., 1999, submitted for publication).

Twenty-five isolates identified by 14 tests appeared different on extended characterisation.

Delayed reactions and interpretation of weak positive and negative reactions were the major differences in the classification of *P. multocida* into species *septica* and *multocida.*

Number of tests used (14 tests) could not adequately separate all species of *P. stomatis, P. canis, P. dagmatis,* and *Pasteurella* taxon 16.

***Table :** Comparative investigations of 25 misclassified strains of Pasteurella. Correct classification, misclassification and major differences observed.*

Extended phenotyping - 79 tests		**Limited phenotyping -14 tests**	**Major differences**
Identification	***No.***	***Misclassified Identification (n)***	
P. multocida.ssp.multocida	3	P. multocida ssp. septica (1) unclassified Pasteurella (2)	Delayed fermentation of sorbitol. Weak fermentation reactions.
P. multocida. ssp. septica	7	P. m. multocida (5), unclassified Pasteurella (2)	Weak and late sorbitol reaction
P. stomatis	2	P. canis (1) P. dagmatis (1)	14 tests insufficient to separate the species
P. dagmatis	2	P. canis	14 tests insufficient to separate the species
Pasteurella Taxon 16	8	P. multocida ssp multocida (3), Unclassified Pasteurella (5). P. dagmatis. (1)	Delayed reactions.14 tests insufficient to separate the species
Unclassified Pasteurella	3	Pasteurella Taxon 16	Weak reactions in many sugars.

Further characterisation of strains of P. multocida showed two different phenotypic clones in chickens and one clone in ducks.

One isolate from chickens was similar to duck isolates in all features. The clones from chickens and ducks differed from dog and cat strains in one or more of the following reactions: glycerol, L (+) arabinose, xylose and trehalose.

Ducks and chickens carrying the same clone of P. multocida ssp. multocida were in the same vicinity, though they were kept in different households.

Discussion

In the present study infectious and noninfectious conditions were shown to cause mortality in the village chickens. Mortalities caused by *E. coli* and fleas were demonstrated in addition to Newcastle disease and injuries. Miscellaneous undiagnosed conditions were also seen. Availability of dead carcasses was limited by lack of reliable transport, and cool storage facilities in the villages, which resulted in decomposition of carcasses. Also number of carcasses obtained per day could not always justify the local contact personnel to bring the carcasses to the University. On the other hand many villagers are normally throwing away dead chicks, whereas moribund adult chickens can be slaughtered for family or animal consumption.

Post-mortem examination results showed that mortalities in seven out of 30 chickens submitted for examination were caused by Newcastle disease. The results support the previous observations that Newcastle disease is prevalent in rural scavenging chickens in Morogoro (Minga et al., 1989; Yongolo, 1996). Age of birds with Newcastle disease ranged from 2 to 4 months. This is within the same age group Thitisak et al. (1988) found in village chickens in Thailand.

E. coli infections, which are so far scantily documented in scavenging village chickens, appeared in four chickens. In the present study three chickens (30%) with colisepticemia were diagnosed from a single clutch with 10 chicks. Unhygienic conditions and poor ventilation in chicken shelters (Mwalusanya, 1998) might favour contamination of eggs or early infection of chicks with the bacterium. Similar management in all families might suggest that *E. coli* is prevalent in many flocks of village chickens. However, because population at risk is mainly chickens less than four weeks, the impact of *E coli* infections is low if compared with diseases like Newcastle disease. However, their significance in immunosuppressed animals should not be underestimated. Shortage of reports of *E. coli* infections in village chickens might also be due to lack of routine post-mortem examination in village chickens in Tanzania (Msanga and Tungaraza, 1985).

Fleas (*E. gallinacea*) have been reported to cause debilitation in indigenous chickens (Cooper, 1967; Msanga and Tungaraza, 1985) and mortality in chicks (Sa'idu et al., 1994). In the current investigation mortalities in five chicks from one flock was caused by fleas. None of the mortalities was, however, caused by endoparasites, which were found in 9/30 of the carcasses investigated. However, poor nutritional condition of worm-infested birds might have contributed to susceptibility to other diseases.

Injuries through management problems appear to cause a significant proportion of mortalities in village chickens. All chickens that died because of injuries were suspected to have been beaten by neighbours. The present findings support the findings of Mwalusanya et al. (1998) that accidents and injuries cause 10% losses in the flocks per annum. This is common during the sowing season if chickens are found in the farms. Seasonal fencing of the birds can significantly reduce mortality caused by management problems (Sonaiya, 1990).

Nine out of 30 cases were not definitely diagnosed; four chickens with renal lesions and three chicks with skin lesions, and two chickens with upper respiratory tract changes. Sa'idu et al. (1994) reported the existence of skin tumours in village chickens. However, details of the lesions were not available. The nature of the skin lesions in the current study remains to be investigated by histopathological techniques. The mortality in these chicks was probably caused by starvation caused by shutting of eyelids caused by thickening skin. Assistants in the villages acknowledged that this skin disease is one of the common diseases that affect chicks in their villages.

The diverse structure of flocks in the village chickens is probably the reason for a biased attention to epidemics that usually affect birds of all age's groups. In village management a flock of chickens is made up of birds of different age groups, including chicks (below 8 weeks), growers (2- 6 months), and adults (after 6 months) in widely different proportions. Consequently, the population at risk to age predisposed diseases is variable and investigation of mortality should address age specific causes. In the present investigation colisepticemia and fleas caused mortality in young chicks, while Newcastle disease affected growers of 2 to 4 months. The present results are comparable to those of Thitisak et al. (1988), who found that infectious coryza and avian pasteurellosis were equally important to Newcastle disease in village chickens. In that study causes of mortality in different age groups were compared. Most farmers visited during this study understand that Newcastle disease is the leading killer of their flocks, but they also know that survival of chicks below eight weeks of age is very low. The wide variety of conditions obtained in the present study implies that application of post-mortem is vital for investigation of causes of mortality in village chickens.

P. multocida has been reported from a wide range of animal hosts (Mutters et al., 1989). Snipes et al. (1988) showed that strains of *P. multocida* from wild animals could cause infection in turkeys. Subsequent investigation by Korbel et al. (1992) demonstrated *P.*

multocida infection in dead feral birds injured by cats. In the present study *P. multocida* has been obtained from dogs and cats kept in contact with domestic poultry. Cat bites are more likely than dog bites to transmit *P. multocida,* when considering the high prevalence shown in cats. Studies of animal bite wounds in human beings have shown that cat wounds are more likely to cause *P. multocida* infection than dog bites (Zbinden et al., 1988). On the other hand village dogs and cats may be transient carriers of *P. multocida* strains from poultry through consumption of dead carcasses. However, determination of persistence of clones of poultry origin in cats and dogs, and consequences of exchange of clones on the pathogenecity remains to be addressed.

The application of 14 tests for identification and typing of *P. multocida* from different animals could not separate isolates from different animals. In addition, using only a limited number of tests for identification resulted in misclassification of species. However, the *P. multocida* strains were separated by extended phenotypical typing. The strains of *P. multocida* from different species could be distinguished despite being similar by using only a few characters. Two phenotypical clones were obtained from chickens, while only one clone was obtained from ducks. One isolate from chick-ens was shown to be similar in all 79 features to isolates from ducks, which were clonal too. This suggests that clones of *P. multocida* can be exchanged between chickens and ducks kept in the same environment. Further characterisation by molecular techniques to confirm the observations is in progress.

Chapter 4

Broiler Management

A broiler is a type of chicken raised specifically for meat production. Modern commercial broilers, typically known as Cornish crosses or Cornish-Rocks are specially bred for large scale, efficient meat production and grow much faster than egg or traditional dual purpose breeds. They are noted for having very fast growth rates, a high feed conversion ratio, and low levels of activity. Broilers often reach a harvest weight of 4-5 pounds dressed in only five weeks.

They have white feathers and yellowish skin. This cross is also favourable for meat production because it lacks the typical "hair" which many breeds have that necessitates singeing after plucking. Both male and female broilers are slaughtered for their meat. In 2003, approximately 42 billion broilers were produced, 80% of which were produced by four companies: Aviagen, Cobb-Vantress, Hubbard Farms, and Hybro.

History

Before the development of modern commercial meat breeds (cows, chickens, etc.) broilers consisted mostly of young male chickens (cockerels) which were culled from farm flocks.

The males were slaughtered for meat and the females (pullets) were kept for egg production. Compared to today, this made chicken meat scarce and expensive compared to eggs, and chicken was a luxury meat. The development of special broiler breeds decoupled the supply of broilers from the demand for eggs. This, along with advances in nutrition and incubation that allowed broilers to be raised year-round, allowed chicken to become a low-cost meat.

Broilers are often called "Rock-Cornish," referring to the adoption of a hybrid variety of chicken produced from a cross of male of a naturally double breasted Cornish strain and a female of a tall, large boned strain of white Plymouth Rocks. This first attempt at a hybrid meat breed was introduced in the 1930s and became dominant in the 1960s. The original cross was plagued by problems of low fertility, slow growth, and disease susceptibility, and modern broilers have gradually become very different from the Cornish x Rock hybrid.

Modern Variants

Access to a special diet of high protein feed delivered via an automated feeding system. This is combined with artificial lighting conditions to stimulate growth and thus the desired body weight is achieved in 4 - 8 weeks, depending on the approximate body weight required by the processing plant. After processing, the poultry is delivered as fresh or frozen chicken to the stores and supermarkets.

Figure : *Five day old broiler strain Cornish-Rock chicks.*

Because of their efficient meat conversion, broiler chickens are also popular in small family farms in rural communities, where a family will raise a small flock of broilers.

Broilers are sometimes reared on a grass range using a method called pastured poultry, as developed by Joel Salatin and promoted by the American Pastured Poultry Producers Association. The term "broiler" is widely known in North America, Australia and England but not elsewhere in the English speaking world. The term "broiler chicken" is very widely used in Pakistan and India, as it was in the former German Democratic Republic and still nowadays in some eastern parts of Germany. The term is also used in Bangladesh, Indonesia, Sweden, Nigeria, Finland, Poland, Turkey and the Balkans.

Broiler Health Issues

Broiler chickens may develop several health issues as a result of selective breeding. Broiler chickens are bred to be very large to produce the most meat per animal. The large chickens cannot stand because their bodies grow too quickly for their legs. Therefore, they may become lame or suffer from broken legs. Broiler chickens are also prone to heart attacks for the same reason, as the heart cannot support blood flow to the large body of the chicken. Another issue with selective breeding is the larger chickens have a more aggressive appetite. The broilers are feed restricted and this leads to behavioural issues in chronically hungry birds.

Broiler chickens may often get joint disorders because their legs cannot bear the heavy bodies. A Swedish study by SLU Skara (Swedish farming university) revealed that only 1/3 of studied broiler chickens that were about to be slaughtered were healthy. Additionally, it is very inactive and as a result is a poor forager, prone to predation, and is generally not suited to small free range homestead flocks.

If the litter in the pen is not properly managed to prevent birds from standing and resting in their feces, painful hock burns and foot ulcerations and blisters can occur. Pastured birds which are rotated frequently typically do not have these issues.

Hatchery

Poultry Hatcheries

Poultry hatcheries produce a majority of the birds consumed in the developed world including chickens, turkeys, ducks and some other minor bird species. It is a multibillion dollar industry, with highly regimented production systems used to maximise bird size versus feed consumed. Birds are produced and maintained under high density, which makes production and harvesting more economical, but can also generate problems such as the spread of pathogens, which can move very quickly through the population when animal densities are high. Poultry generally start with naturally (chickens) or artificially (turkeys) inseminated hens that lay eggs; the eggs are cleaned and shells are checked for soundness before putting them in the incubators. The incubators control temperature and humidity, and turn the eggs until they hatch. Generally large numbers are produced at one time so the resulting birds are uniform in size and can be harvested at the same time. Once the eggs hatch and the chicks are a few days old, they are often vaccinated, beak-trimmed and or toe-clipped; this involves the removal of half of the top beak and the clipping of the toe ends. This is done to prevent the birds from harming each other while they are living in close proximity to each other. After these procedures, they are moved to enclosed buildings to be raised until harvest.

For chickens bred as (egg) layers, only the female chicks are considered to be of value; in excess of 100,000 male chicks can be dropped into an industrial "grinder" and disposed of in a single day.

Bantam (Poultry)

A bantam is a small variety of poultry, especially chickens. Etymologically, the name *bantam* is derived from the city of Bantam - currently known as "Banten Province" or previously "Banten Residency" - once a major seaport, in Indonesia. European sailors restocking on live fowl for sea journeys found the small native breeds of chicken in Southeast Asia to be useful, and any such small poultry came to be known as a *bantam*.

Most large chicken breeds have a bantam counterpart, sometimes referred to as a *miniature*. Miniatures are usually one-fifth to one-quarter the size of the standard breed, but they are expected to exhibit all of the standard breed's characteristics.

Characteristics

Bantams are suitable for smaller backyards as they do not need as much space as other breeds. Bantam hens are also used as laying hens, although Bantam eggs are only about one-half to one-third the size of a regular hen egg. The Bantam chicken eats the same foods as a normal chicken, chickens in the wild eat more insects and vegetation than grains. In commercial situations they are fed grain based foods because this is convenient and efficient for the producer. Bantams have become increasingly popular as pets as well as for show purposes because they are smaller and have more varied and exotic colours and feather patterns than other chickens. Breeds such as the Sebright, Dutch, and Pekin are particularly popular show birds, and true bantams.

In contrast, the Bantam rooster is famous in rural areas throughout the United Kingdom and the United States for its aggressive, "puffed-up" disposition that can be comedic in stature. It is often called a "Banty" in the rural United States.

Many bantam hens are renowned for hatching and brooding purpose. They are very protective mothers and will attack anything that gets near their young.

The Bantam chicken is considered faster and "spunkier" than their larger counterparts. In 1954, the Trinity College basketball team had named their team after the Bantam chicken. To this day, Trinity College still calls their basketball team the "Trinity Bantams".

Old English bantam roosters were commonly used for fighting in Europe. They were smaller and faster than normal roosters that were used previously.

Bantams do have a higher mortality rate when they are kept as backyard pets. They are easy targets for hawks, cats, foxes, or any other small predator. The average backyard free range bantam lives 1-3 years.

True Bantams

- A true bantam has no large counterpart, and is naturally small. Such birds are often popular for show purposes.
- Birds designated as true bantams include:
- Belgian Bearded d'Anvers
- Belgian Rumpless d'Anvers
- Belgian Bearded d'Uccle
- Belgian Rumpless d'Uccle
- Belgian Bearded de Watermael
- Booted
- Dutch
- Japanese
- Nankin
- Pekin
- Rosecomb
- Sebright
- Tuzo

Chickens, turkeys, ducks, and geese are of primary importance, while guinea fowl and squabs are chiefly of local interest.

Chickens

Humans first domesticated chickens of Indian origin for the purpose of cockfighting in Asia, Africa, and Europe. Very little formal attention was given to egg or meat production. Cockfighting was outlawed in England in 1849 and in most other countries thereafter. Exotic breeds and new standard breeds of chickens proliferated in the years to follow, and poultry shows became very popular. From 1890 to 1920 chicken raisers stressed egg and meat production, and commercial hatcheries became important after 1920.

Breeds

The breeds of chickens are generally classified as American, Mediterranean, English, and Asiatic. The American breeds of importance today are the Plymouth Rock, the Wyandotte, the Rhode Island Red, and the New Hampshire. The Barred Plymouth Rock, developed in 1865 by crossing the Dominique with the Black Cochin, has grayish-white plumage crossed with dark bars. It has good size and meat quality and is a good layer. The White Plymouth Rock, a variety of the Barred Plymouth Rock, has white plumage and is raised for its meat. Both varieties lay brown eggs. The Wyandotte, developed in 1870 from five or more strains and breeds, has eight varieties and is characterised by a plump body, excellent meat, and good egg production. Only the white strain is of any significance today because it is used in broiler crosses where its white plumage, quality of flesh, and rapid growth are highly desirable.

An American breed, the Rhode Island Red, developed in 1857 from Red Malay game fowl crossed with reddish-coloured Shanghais—with some brown Leghorn, Cornish, Wyandotte, and Brahma blood—is good for meat production and is one of the top meat breeds for the production of eggs. It has brilliant red feathers and lays brown eggs.

The New Hampshire, developed in the U.S. in 1930 from Rhode Island Red stock, is a meaty, early maturing breed with light-red feathers and lays large brown eggs. The only Mediterranean breed of importance today is the Leghorn. This breed, originated in Italy, has 12 varieties, the single-comb White Leghorn being more popular than all of the other types combined. This breed, the leading egg producer of the world, lays white eggs and is kept in large numbers in England, Canada, Australia, and the U.S. The White Minorca, a second Mediterranean breed, is often used in crossbreeding for egg production.

The only English breed of modern significance is the Cornish, a compact and heavily meated bird used in crossbreeding programs for broiler production. It is a poor producer of eggs, however. The only Asiatic breed of significance today, the Brahma, which originated in India, has three varieties, the light Brahma being preferred because of its size.

Chicken breeding is an outstanding example of the application of basic genetic principles of inbreeding, linebreeding, and crossbreeding, as well as of intensive mass selection to effect faster and cheaper gains in broilers and maximum egg production for the egg-laying strains. Maximum use of heterosis, or hybrid vigour, through incrosses and crossbreeding has been made. Crossbreeding for egg production has used the single-comb White Leghorn, the Rhode Island Red, the New Hampshire, the Barred Plymouth Rock, the White Plymouth Rock, the Black Australorp, and the White Minorca. Crossbreeding for broiler production has used the White Plymouth Rock or New Hampshire crossed with White or Silver Cornish or incrosses utilising widely diverse inbred strains within a single breed. Rapid and efficient weight gains, and high quality, plump, meaty carcasses have been achieved thereby.

The male sperm lives in the hen's oviduct for two to three weeks. Eggs are fertilised within 24 hours after mating. Yolks originate in the ovary and grow to about 1.6 inches (4.0 centimetres) in diametre, after which they are released into the oviduct, where the thick white and two shell membranes are added. The egg then moves into the uterus where the thin white and the shell are added. This process requires a total of 24 hours per egg. The hatching of fertilised eggs requires 21 days, with the heavy breeds requiring a few more hours and the lighter breeds slightly fewer. Ideal hatching temperature approximates 100° F (38° C) with control of air flow, humidity, oxygen, and carbon dioxide being essential. Standardised egg-laying tests and official random sample tests have been used for many years to measure actual productivity.

Feeding

Chicken feeding is a highly perfected science that ensures a maximum intake of energy for growth and fat production. High quality and well-balanced protein sources produce a maximum amount of muscle, organ, skin, and feather growth.

The essential minerals produce bones and eggs; 3 to 4 percent of the live bird being composed of minerals and 10 percent of the egg. Calcium, phosphorus, sodium, chlorine, potassium, sulfur, manganese, iron, copper, cobalt, magnesium, and zinc are all required. Vitamins

A, C, D, E and K and all 12 of the B vitamins are also required. Water is essential, and antibiotics are almost universally used to stimulate appetite, control harmful bacteria, and prevent disease. Modern rations produce a pound of broiler on about two pounds (0.9 kilograms) of feed and a dozen eggs from 412 pounds (2.0 kilograms) of feed.

Management

Among the world's agricultural industries, meat chicken breeding in the U.S. is one of the most advanced. It is presently considered the model for other animal industries, the broiler industry leading the way in advanced agricultural technology and efficiency. Intensive nutritional research and application, highly improved breeding stock, intelligent management, and scientific disease control have gone into the effort to give a modern broiler of uniformly high quality produced at ever-lower cost.

Today, one person can care for 25,000 to 50,000 broilers that reach market weight in three months' time, giving an annual output of from 100,000 to 200,000 broilers. A modern broiler chick gains over 43 times its initial weight in an eight-week period. Aggressive marketing methods increased the per capita consumption of broilers more than fivefold in the three decades beginning in 1950, with further substantial increases predicted for the future.

Less than half as much feed is now required to produce a pound of broiler meat as was needed in 1940. While per capita consumption of eggs has declined, the feed requirement per dozen eggs is only slightly more than half as high as it was in the early 1900s. Annual egg production per hen has increased from 104 to 244 since 1910.

A carefully controlled environment that avoids crowding, chilling, overheating, or frightening is almost universal in chicken raising. Cannibalism, which expresses itself as toe picking, feather picking, and tail picking, is controlled by debeaking at one day of age and by other management practices. The feeding, watering, egg gathering, and cleaning operations are highly mechanised. More than 90 percent of the 4,200,000,000 chicks hatched per year in the early 1980s were The vast majority of chicks hatched each year are used for broiler production and the remainder for egg production. In egg production feed represents more than two-thirds of the cost. Pullet (immature hen) flocks predominate. Hens are usually housed in wire cages with two or three hens per cage and three or four tiers of cages superposed to save space. Cages for laying hens have been found to increase production, lower mortality, reduce cannibalism, lower feeding requirements, reduce diseases and parasites, improve culling, and reduce both space and labour requirements.

Other Poultry

- These include turkeys, ducks, geese, guinea fowl, and squabs.
- Turkey production.

After World War II turkey production became highly specialised, with larger flocks predominating. Turkeys are raised in great numbers in Canada where their ancestors still live wild, as also in some parts of the U.S. Broad Breasted Bronze, Broad Breasted White, and White Holland are the most popular of the larger breeds, representing nearly three-fourths of the total production. The Beltsville Small White is the most popular of the smaller breeds and composes the bulk of the remaining 25 percent. At 24 weeks of age the toms are 50 percent heavier than the hens. In breeding flocks, one tom is required per eight or 10 hens.

Tremendous improvements both in breeding and nutrition have been made in this century. Since 1910, the amount of feed required to produce a pound of turkey meat has fallen 40 percent, while the time required has been reduced 25 percent. Fifty to 80 pounds (23–36 kilograms) of feed will produce a turkey for market weight with from 212 to 3 pounds required per pound of gain on full-size turkeys, and 212 to 234 pounds (1.1–1.2 kilograms) of feed per pound (0.45 kilograms) of gain for turkey broilers, which are marketed at from 12 to 15 weeks of age. Turkey poults are hard to start on feed. One method is to dip their beaks in water and then in feed. Another is to light the feed troughs very brightly and to use oatmeal or ground yellow corn sprinkled on top of the feed. Turkeys are given range, or open land, and automatic waterers, self-feeders, range shelters, heavy fencing, and rotated pastures are used. Successful marketing techniques have increased turkey consumption; *e.g.*, in the U.S., per capita consumption from 1930/34 to 1980 rose 500 percent.

Duck and Goose Production

Duck raising is practised on a limited scale in nearly all countries, for the most part as a small-farm enterprise. The flocks once kept in England are much reduced, the demand for eggs being greatly lessened, though a limited market still exists. Khaki Campbell and Indian Runner ducks are prolific layers, each averaging 300 eggs per year. In Indonesia, where the labour supply is large, duck herders take a flock of ducks to the high country during the warmer seasons and work their way down the mountainsides to the lowlands.

Ducks are easily transported, can be raised in close confinement, and convert some waste products and scattered grain (*e.g.*, by gleaning rice fields) to nutritious and very desirable eggs and meat. In developed countries, commercial plants have been built exclusively for duck meat production; an example is the large duckling industry of Long Island, New York. There are also local industries in The Netherlands and England, the favourite breed in England being the Aylesbury. This breed has white flesh and can reach eight pounds (3.6 kilograms) in eight weeks. The U.S. favourite is the Pekin duck, which is slightly smaller than the Aylesbury and yellow-fleshed.

Goose raising is a minor farm enterprise in practically all countries, but in Germany, Austria, some eastern European countries (notably Poland), parts of France, and locally elsewhere, there is important commercial goose production. The two outstanding meat breeds are the Toulouse, predominantly gray in colour, and the Embden (or Emden), which is white. Geese do not appear to have attracted the attention of geneticists on the same scale as the meat chicken and the turkey, and no change in the goose industry comparable to that in the others has occurred or seems to be in prospect. In some commercial plants, geese are fattened by a special process resulting in a considerable enlargement of their livers, which are sold as a delicacy, pâté de foie gras.

Guinea Fowl and Squabs

Guinea fowl are raised as a sideline on a few farms in many countries, and eaten as gourmet items. In Italy there is a fairly extensive industry.

There the birds are raised in yards with open-fronted shelters. In England, guinea fowl are marketed at 16–18 weeks of age and in the U.S. at about 10–12 weeks. The market weight is usually about 212–312 pounds, but food conversion is poor.

Pigeons are raised not only as messengers and for sport but also for the meat of their squabs (nestlings), also a gourmet item. Squab production, carried on locally, is rare in most countries with established poultry industries.

Poultry Diseases

Poultry are quite susceptible to a number of diseases; some of the more common are fowl typhoid, pullorum, fowl cholera, chronic respiratory disease, infectious sinusitis, infectious coryza, avian infectious hepatitis, infectious synovitis, bluecomb, Newcastle disease, fowl pox, avian leukosis complex, coccidiosis, blackhead, infectious laryngotracheitis, infectious bronchitis, and erysipelas.

Strict sanitary precautions, the intelligent use of antibiotics and vaccines, and the widespread use of cages for layers and confinement rearing for broilers have made it possible to effect satisfactory disease control.

Parasitic diseases of poultry, including hexamitiasis of turkeys, are caused by roundworms, tapeworms, lice, and mites. Again, modern methods of sanitation, prevention, and treatment provide excellent control.

Poussin (Chicken)

In Commonwealth countries, poussin is a butcher's term for a young chicken, less than 28 days old at slaughter and usually weighing 400-450 grammes but not above 750g. It is sometimes also called spring chicken, although the term spring chicken usually refers to chickens weighing 750-850g.

In the United States of America, *poussin* is an alternative name for a small-sized cross-breed chicken called Rock Cornish game hen, developed in the late 1950s, which is twice as old and twice as large as the typical British poussin.

Asil (Chicken)

Figure : *A Reza Asil cock and two hens*

The Asil or Aseel is a breed of chicken originating from South Punjab/Sindh area of Pakistan and India. Similar fowl are found throughout Southeast Asia and have names like Shamo, Taiwan, etc.

Similar to Asils are Sadal (called Malay in Europe). This is a very large breed of chicken from Pakistan and India. They have longer legs with thin thighs and little wattles with pea-combs. The difference between the two is that Sadal are not game (do not fight), and Asil do. Asils were first used for cock fighting. Aseel is noted for its pugnacity. The chicks often fight when they are just a few weeks old and mature roosters will fight to the death. Hens can also be very aggressive towards each other.

Towards humans Asil are generally very tame and trusting. There are anecdotes where they have come to their keepers for other things than food, for example to get the keeper to open the door to the coop so they can get to roost.

The hens are not good layers, but are excellent sitters. Laying depends on the Asil variety, the small Asil are known to be very poor layers, sometimes laying just 6 eggs a year, whereas larger Asil can lay around 40 eggs a year.

In the U.S., the breed is listed as Critical by the American Livestock Breeds Conservancy. Aseel breed is found in almost all states of India, but abundant in Andhra Pradesh.

Breed Standard

The Asil has a distinctive upright stance, drooping tail, and well-defined musculature. Asils are good fighters The colour ranges from white to black with black breasted red being the most common.

Asil Head

The ideal Asil head is more or less round-shaped and broad, the eyes pearl white and protected by protuberant eyebrows and cheekbones, a small low set pea or walnut comb (except the single-combed Bihangham variety) with a relative short and thick beak. The colour of the face is generally red. Asil with a dark-coloured face are seen on South Indian Asil. Wattles should be absent. Only rudimentary presence is allowed.

Asil Comb Types

Asil show a variation in comb types and beak shapes. In generally we can say that North Indian Asil types such as the Reza Asil have (triple) pea combs only. The South Indian varieties such as the Malay and the Madras Asil show (triple) pea combs as well as walnut combs. Comb colour is red. Some varieties such as the Bihangham carry a single comb due to a throw back to the red jungle fowl, however these type of Asil are uncommon even in their homeland.

Wattles

Wattles should be small to absent, absent is preferred.

Asil Beak Types

The smaller the comb the better on Asil. North Indian type Asil have (triple) pea combs and a fairly large beak with the shape similar to an eagle. The birds from Southern India generally show a short but massive triangle shaped beak. Beak colour is ivory white.

Asil Eyes

Original asils have blue colour eyes when young and may turn pearl-white when grown up. Red or orange-coloured eyes are rear depending on the breed. Pale yellow-coloured eyes can be seen in young birds which lighten with age into a pearl-white colour.

Sometimes pearl-white coloured eyes are seen showing tiny blood veins running in the eyes, so called “bloodshot” eyes. In some areas these are regarded as a sign of vitality.

Asil Legs

Leg colour is ivory white. Black legs are acceptable for black colour Asils. The other main leg colours within the Asil breed are pale,grey and white, black, though they are considered inferior.

The dark-coloured leg colours are generally seen on the South Indian Asil. Some Asil show very rough scales pointing a little bit outwards. One the ways you recognise an asil (aseel) is by the legs if they are yellow it means there not pure aseel. what they could be is a shamo, taiwaan or other game birds.

Asil Body Description

The Asil should have broad shoulders and wings are carried against the body. The body of an Asil is very muscular but also compact.

Varieties

There are many varieties of Asil, some are standardised for shows such as the Reza Asil in the UK, some are simply named after the area where they are bred such as the Mianwali Asil from Pakistan or the colour, red/wheaten Asil are generally known as “Sonatol”.

There are also hen-feathered Asil knows as "Madaroo" these are found in various colours, but the cocks come with feathers in hen colour, don't have sickle feathers in the tails and miss the large hanging feathers on the saddle. This variety is very rare.

Figure : *Portrait of Kulang Asil rooster head*

Asil with feather beards under their beaks known as "muffed" and with tufts on the top of their heads known as "tasseled" are also seen, but are very rare especially outside India/Pakistan.

Bhaingam Asil variety have a have a large single comb but confirm to all the other Asil standards.

Broadly speaking, Asil in Europe are categorised and shown under these three types: Madras asil Madras asils are very big and muscular. They can get up to 32 inches the main colours are black, red, grey, blue and green.

Reza Asil

Height: Up to 50 cms tall. Weight: Maximum weight for the hens is 1.8 kg, max weight for the cocks is 2.7 kg.

This type is standardised by the Asian Hardfeather Society in the UK and is seen at shows throughout the UK, but is quite rare.

This group of Asil reached worldwide popularity due to books and articles written by gamefowl experts such as Herbert Atkinson, Siran and Paul Deraniyagala from Sri Lanka and Carlos Finsterbusch from Chile. The Reza Asil family according the old (Western) gamefowl literature is subdivided into following strains: (Amir) Ghan (Dark-Red),

Sonatol(Light-Red), (Siyah) Rampur(Black), Kalkatiya (Kaptan) (Speckled-Reds) and Jawa(Duckwing). All these strains are identified by their specific colour, these colours do not necessarily correspond with the area where the birds come from.

In colonial times other colours such as whites, spangles, golden etc. were regarded as inferior. At present day the "classic" strains and names given mentioned by Atkinson are more or less forgotten. The native people in India, Pakistan, Bangladesh and Sri Lanka only know the Reza-type Asil by their local names.

Kulang Asil

Height: Up to 75 cms Tall. Weight: 5 to 7 kg.

The large Asil are divided into sub-varieties : North Indian, South Indian and Madras type. The North and South Indian varieties don't differ much. Only type of comb, shape of the beak and body shape are different. For example : Northern type = slender, Southern type = heavier build. The Madras Asil however is significantly different. They have a lower station, are heavier build and stronger boned. These birds often come in a bluish colour. This variety is found in the deep south of India, the Tamil Nadu state.

Sindhi Breed

It is one of the tallest and biggest breed of Asils. Main colours are red and blue. They are mostly fought in the Sindh area of Pakistan. These aseels have good endurance and usually their fights last longer than Mianwali Aseels. With the arrival of Mianwali they have started to disappear from the fights.

Mianwali Breed

This breed is mainly found in Mianwali district of Pakistan. However since its arrival, this breed has risen to popularity in Pakistan, currently the primary game breed used in the pits also preferred by gamblers. It is smaller compared to Sindhi aseels weighing between 1.5 to 3.5 kg depending on the preference of breeders. It is much faster and a better head hitter usually comes in small to medium height.

A good Mianwali aseel should kill its opponent with in few minutes. They have been known to kill bigger roosters because of their speed and accuracy. They come in various colours such as Java (duckwing), Lakha (reddish), black and various others depending on the combination used in breeding. Very hard and a brave fighter with attitude to inspire, excellent in naked heels and metal spurs.

There are many sub breeds of this breed owing to the combination used in breeding. A good tested Mianwali rooster would usually have offspring of a similar quality.

Typical description would be small curved beak, strong joints, pearl/white/yellow eye colour, short crow, small comb and do not have heavy body structure. May look smaller than other breeds but is excellent spurer.

Amroha

This is a rare breed of Aseel used in Pakistan and India. Very few of these roosters exist in their pure form. They are known to be small to medium like Mianwali. It is also known that they are champion of naked heel fighting. In simple, it is a fantasy of most aseel breeders in Pakistan.

Bantam Asil

Weight: Up to 0.75 kg.

Bantam Asil have been created at the end of the 19th century by an English breeder named William Flamank Entwisle. The breed got very popular after its creation but after a couple of decades interest in this variety slowly died out.

Until the beginning of the 1980s nothing was heard about these little Asil. A Belgian breeder named Willy Coppens created them again using Shamo (chicken), Indian Game and Reza Asil.

The breed was also introduced again in Holland and United Kingdom. At present day Bantam Asil are quite popular and they are bred in various colours.

Brahma (Chicken)

Brahmas are an Asiatic breed of chicken, originating in the Brahmaputra region in India where they were known as "Gray Chittagongs." Their heritage is unclear, but they are believed to be closely related to the Jungle Fowl (Gallus Gigantus) and the Cochin. The first Brahmas were brought to the U.S. from British India in 1846, and were used as a utility fowl for their edibility and generous egg laying and hardiness even during the winter months, although today they are kept mainly for ornamental purposes as selection for utility has taken a back seat to selection for appearance.

Some of the earliest imports to the U.S. reached weights of nearly 14 pounds, but rarely is such massive size seen today: standard weight for a cock is 11 pounds; hens are 8.5 pounds. By the 1870s Brahmas had become so popular that they were admitted into the American Poultry Association's Standard of Perfection.

Temperament

Brahmas are calm, friendly birds that make good pets or exhibition fowl. Males are calm and generally not aggressive towards humans. They are not skittish or easily scared, making them a popular choice for families with children.

Due to their docile demeanour, Brahmas can be easily trained so that they can be handled by almost anyone. They should be hand trained when young because their large size makes them difficult to control in the early stages of training if they are full grown.

Appearance

Brahmas are massive in appearance, in part due to profuse, loose feathering and feathered legs and toes. Approximate weights:

Cock - 12 pounds (5.443 kg)

Cockerel - 10 pounds (4.536 kg)

Hen - 9 pounds

Pullet - 8 pounds

Recognised Varieties

The American Standard of Perfection recognises three Brahma varieties: light, dark, and buff.

The light Brahma has a base colour of white, with black hackles edged in white and a black tail. The cocks' saddle feathers in a light Brahma are striped with black.

The dark Brahma has the most notable difference between cock and hen. The hen has a dark gray and black pencilled coloration with the same hackle as the light whereas the cock has black and white hackles and saddle feathers, and a black base and tail.

The wings of a dark Brahma are white-shouldered and the primary feathers (remiges) are edged with white. Buff Brahmas have the same pattern of black as light Brahmas, except with a golden buff base colour instead of white.

In Australia Brahma Breeders are creating more colours and along with the accepted American varieties - Light, dark, and buff the Australian Poultry Association have accepted black, blue, partridge, crele and even barred varieties of Brahma.

Intensive Chicken Farming

In egg-producing farms, birds are typically housed in rows of battery cages. Environmental conditions are automatically controlled, including light duration, which mimics summer daylength. This stimulates the birds to continue to lay eggs all year round. Normally, significant egg production only occurs in the warmer months. Critics argue that year-round egg production stresses the birds more than normal seasonal production.

Meat chickens, commonly called broilers, are floor-raised on litter such as wood shavings or rice hulls, indoors in climate-controlled housing. Poultry producers routinely use nationally approved medications, such as antibiotics, in feed or drinking water, to treat disease or to prevent disease outbreaks arising from overcrowded or unsanitary conditions.

In the U.S., the national organisation overseeing chicken production is the Food and Drug Administration (F.D.A.). Some F.D.A.-approved medications are also approved for improved feed utilisation. In the U.S., federal law prohibits the use of hormones or steroids in poultry production.

In egg-producing farms, cages allow for more birds per unit area, and this allows for greater productivity and lower space and food costs, with more efforts put into egg-laying. In the U.S., for example, the current recommendation by the United Egg Producers is 67 to 86 in^2 (430 to 560 cm^2) per bird, which is about 9 inches by 9 inches. Modern poultry farming is very efficient and allows meat and eggs to be available to the consumer in all seasons at a lower cost than free range production, and the poultry have no exposure to predators.

The cage environment of egg producing does not permit birds to roam. The closeness of chickens to one another frequently causes cannibalism. Cannibalism is controlled by debeaking (removing a portion of the bird's beak with a hot blade so the bird cannot effectively peck). Another condition that can occur in prolific egg laying breeds is osteoporosis.

This is caused from year-round rather than seasonal egg production, and results in chickens whose legs cannot support them and so can no longer walk. During egg production, large amounts of calcium are transferred from bones to create eggshell. Although dietary calcium levels are adequate, absorption of dietary calcium is not always sufficient, given the intensity of production, to fully replenish bone calcium.

Under intensive farming methods, a meat chicken will live less than six weeks before slaughter. This is half the time it would take traditionally. This compares with free-range chickens which will usually be slaughtered at 8 weeks, and organic ones at around 12 weeks.

In intensive broiler sheds, the air can become highly polluted with ammonia from the droppings. This can damage the chickens' eyes and respiratory systems and can cause painful burns on their legs (called hock burns) and feet. Chickens bred for fast growth have a high rate of leg deformities because they cannot support their increased body weight. Because they cannot move easily, the chickens are not able to adjust their environment to avoid heat, cold or dirt as they would in natural conditions. The added weight and overcrowding also puts a strain on their hearts and lungs. In the U.K., up to 19 million chickens die in their sheds from heart failure each year.

Chapter 5

Egg Drop Syndrome

Egg drop syndrome (EDS) is characterised by production of soft-shelled and shell-less eggs in apparently healthy birds. It has been recognised worldwide, except in the USA.

Etiology

The causal adenovirus is widely distributed in both wild and domestic ducks, geese, coots, and grebes. Antibody has also been detected in herring gulls, owls, storks, and swans. The adenovirus group antigen cannot be demonstrated by conventional means, and EDS virus also differs from other avian adenoviruses by strongly agglutinating avian RBC. The virus achieves high titres in embryonating eggs or in cell cultures of duck or goose origin. It replicates well in chick kidney or chick-embryo liver cells and to a lesser degree in chick-embryo fibroblasts. It does not grow in embryonating chick eggs or in mammalian cells.

The resistant virus has 1 serotype but at least 3 genotypes: 1 associated with classical EDS, 1 with ducks in the UK, and 1 with EDS in Australia.

Epidemiology

The natural hosts for EDS virus are ducks and geese, and the disease has been described in Japanese quail (Coturnix coturnix *japonica*). Three types of disease are recognised in chickens. Classical EDS probably was due to contamination of a vaccine for Marek's disease grown in duck-embryo fibroblasts and subsequent adaptation of the virus to chickens. Basic breeding stock was infected, and the virus was transmitted vertically through the egg. The virus often remained latent until the chick reached sexual maturity, when it was excreted in the eggs and droppings to infect susceptible contacts. Because the virus is vertically transmitted and is reactivated around peak egg production, there was an apparent breed and age susceptibility. However, all ages and breeds of chickens are susceptible, although the disease tends to be most severe in heavy broiler-breeders or brown egg producers.

Arising from the classical form, endemic EDS has been reported in many areas and is usually seen in commercial egg producers. Flocks become infected at any stage in lay. Contaminated egg collection trays are one of the main forms of horizontal transmission, and outbreaks are often associated with a common egg-packing station.

Rare, sporadic EDS has been recognised in isolated flocks. It appears to be due either to contact with domestic ducks or geese or, more often, to water contaminated with wildfowl droppings. The risk is that these introductions could become endemic.

The main method of horizontal spread is through contaminated eggs; droppings also are infective. Humans and contaminated fomites such as crates or trucks can spread virus, which also can be transmitted by needles when vaccinating and drawing blood. Insect transmission is possible but not proved.

Pathogenesis

After horizontal or experimental infection, the virus grows to low titres in the nasal mucosa. This is followed by viraemia, virus replication in lymphoid tissue, and then massive replication for ~8 days in the oviduct, especially in the pouch shell gland region. Changes in the eggshell occur coincidentally. Both the exterior and interior of eggs produced between 8 and ~18 days after infection contain virus. A copious exudate in the lumen of the oviduct is rich in virus, and this contaminates the droppings. Unlike other fowl adenoviruses, there is little, if any, growth in the epithelial cells of the intestine.

Chicks hatched from infected eggs may excrete virus and develop antibody. More often, the virus remains latent, and antibody does not develop until the bird starts to lay, at which time the virus reactivates and grows in the oviduct, repeating the cycle.

Clinical Findings

In flocks without antibody, the first sign is loss of colour in pigmented eggs, quickly followed by soft-shelled and shell-less eggs. Diarrhea and a transient dullness may be seen before the eggshell changes. Birds tend to eat the shell-less eggs, which therefore may be missed unless a search is made for the membranes. Egg production falls 10-40% mainly because of the shell-less eggs. In flocks in which there has been some spread of virus and some of the birds have antibody (usually 10-20%), the condition is seen as a failure to achieve predicted production targets; careful examination shows that these flocks are experiencing a series of small EDS episodes. Birds with antibody slow the spread of virus.

There is no effect on fertility or hatchability of those eggs suitable for setting.

Lesions

The major pathologic changes are seen in the pouch shell gland. Surface epithelial cells develop intranuclear inclusion bodies and degenerate; they are replaced by squamous, cuboidal, or undifferentiated columnar cells. There is moderate to severe inflammatory infiltration of the mucosa.

Diagnosis

In classical EDS, the combination of poor eggshell quality at peak production in healthy birds is almost diagnostic. With endemic or sporadic EDS, disease can be seen in laying birds of any age. In cage units, spread can be slow, and the clinical signs, may be overlooked or perceived as a small depression (2-4%) of egg yield.

EDS can be distinguished from Newcastle disease (*Newcastle Disease*) and influenza virus infections (*Avian Influenza: Introduction*) by the absence of illness, and from infectious bronchitis (*Infectious Bronchitis: Introduction*) by the eggshell changes that occur at or just before the drop in egg production and by the absence of ridges and malformed eggs sometimes seen in infectious bronchitis.

EDS should be suspected whenever peak egg production parametres are not met; however, the observation of clinical signs alone does not provide enough reliable information for diagnosis. The virus can be isolated by inoculating embryonating duck eggs or duck- or chick-embryo liver cell cultures. It is important to select birds producing abnormal eggs, but this can be difficult, especially if the birds are on litter. An easier method is to feed affected eggs to antibody-free hens. Virus isolation from the pouch shell gland of these hens can be attempted when the first abnormal eggs are produced.

The hemagglutination inhibition test (high levels of hemagglutinins are produced) using fowl RBC or the ELISA test are the serologic and diagnostic tests of choice. In addition, the serum neutralisation test can be used for confirmation. The double immunodiffusion test also has been used. If one adenovirus has been isolated, restriction endonuclease analysis can be used to classify the virus as EDS. When selecting birds for diagnosis, especially in cage units, it is important to bleed only birds that have produced affected eggs.

Control

There is no treatment. The classical form has been eradicated from primary breeders. Washing and disinfecting plastic egg trays before use can control the endemic form. The sporadic form can be prevented by separating chickens from other birds, especially waterfowl. General sanitary precautions are indicated, and potentially contaminated water should be chlorinated before use.

Inactivated vaccines with oil adjuvant are available and, if properly made, control the disease. They reduce but do not prevent virus shedding. These vaccines are given during the growing phase, usually at 14-18 wk, and can be combined with other vaccines such as for Newcastle disease. Sentinel birds are frequently placed along with vaccinated chickens to detect the presence of virus in the flock. Sentinel chickens will become serologically positive on hemagglutination inhibition test.

Avian Pneumovirus

Turkey Rhinotracheitis, Avian Rhinotracheitis, Swollen Hea

Avian pneumoviruses have been implicated in the upper respiratory tract disease of turkeys and chickens known as turkey rhinotracheitis. The virus has also been associated with swollen head syndrome of chickens. First described in south Africa in the late 1970s, avian pneumovirus soon appeared in Europe and the Middle East. The disease has now been reported from all the major poultry-producing areas in the world except for Australasia. The virus has also been detected in pheasants and guinea fowl, and serologic evidence suggests other avian species are susceptible. Recent studies have indicated an antigenic relationship with a newly discovered human pneumovirus isolated from children with respiratory tract disease.

Etiology

Avian pneumoviruses are members of the subfamily Pneumovirinae, belonging to the family Paramyxoviridae. The subfamily consists of 2 genera: Pneumovirus, consisting of mammalian respiratory syncytial viruses and mouse pneumovirus, and Metapneumovirus, in which avian pneumoviruses are placed. Based on observed differences following sequence analysis of the viral genes, at least 3 subtypes of the virus have been described (A, B, C). Subtype C viruses appear to be the only subtype found in North America and have not been reported in other parts of the world.

Transmission and Epidemiology

Following infection, the virus is shed from the nares and trachea but not in the feces. After initial introduction of the virus, the disease spreads rapidly within a geographic area or country. The methods by which the virus is spread are unclear and often unpredictable. Direct contact attributed to movement of infected birds, personnel, equipment, and vehicles have all been implicated, while airborne transmission has also been reported to occur. There is no published evidence of vertical transmission via the egg, even though the virus has, on occasion, been detected in the reproductive tract of laying birds. Persistence of the virus in turkeys and chickens has not been demonstrated and, following experimental infection, virus was detected for 6-7 days only after inoculation. Wild birds have been implicated in the spread of avian pneumovirus, particularly waterfowl and gulls.

Clinical Findings

Clinical signs in turkey poults include snicking, rales, sneezing, nasal discharge, foamy conjunctivitis, swollen infraorbital sinuses, and submandibular edema. Coughing and head shaking are frequently observed in older poults. In laying birds, egg production may drop up to 70% with an increased incidence of poor shell quality and peritonitis. Coughing associated with lower respiratory tract involvement may lead to prolapses of the uterus in laying turkeys.

Morbidity in birds of all ages is usually described as up to 100% with mortality ranging from 0.4% to as high as 50%, particularly in fully susceptible young poults. Secondary pathogens and management factors significantly influence the levels of morbidity and mortality.

Infection in chickens and pheasants is less clearly defined and may not always be associated with clinical signs. Avian pneumovirus is associated with swollen head syndrome in chickens. This condition is characterised by swelling of the peri and infraorbital sinuses, torticollis, cerebral disorientation, and opisthotonos. Typically, <4% of the flock is affected, although respiratory signs may be widespread. Mortality is rarely >2%. In broiler breeders and commercial layers, egg production and quality are frequently affected. Evidence suggests that infectious bronchitis virus (*Infectious Bronchitis: Introduction*) and Escherichia coli may also be associated with swollen head syndrome.

Lesions

In turkeys, excess mucus found in the nares, sinuses, and trachea is clear at first but rapidly becomes mucopurulent. When bacteria are involved, typical lesions of colisepticemia are frequently seen in various organs. The oviducts of affected breeders may contain inspissated albumen and solid yolk. Egg peritonitis may be associated with oviduct regression. Microscopic examination of the upper respiratory tract 1-2 days after infection reveals localised lesions, including loss of cilia, increased glandular activity, congestion, and mild mononuclear infiltration of the submucosa. Inflammatory infiltration of the submucosa can be observed with some mild lesions in the trachea between days 3 and 5. Similar but milder lesions can be observed in affected chickens.

Diagnosis

Taking samples from the upper respiratory tract of birds in the very early stages of the disease is extremely important when attempting virus isolation. Early investigators found that tracheal organ cultures prepared from turkey or chicken embryos, or 1 to 2 day old chicks, were the most sensitive for primary isolation of avian pneumoviruses. Ciliostasis may occur within 7 days of inoculation or on passage. The virus has also been isolated following the inoculation of 6 to 8 day old embryonated chicken or turkey eggs via the yolk sac route and identified by electron microscopy, virus neutralisation, and molecular techniques. Cell cultures have not proved successful for the primary isolation of the virus. However, once the virus has been isolated and adapted in the systems above, it will grow in a variety of avian and mammalian cultures.

PCR tests have been developed and are widely used to detect the virus in clinical material, particularly respiratory swabs. Some PCR tests have been constructed so that the subtype as well as the identity of virus can be determined from the clinical sample. Antigen detection tests have also been developed, including immunofluorescence and immunoperoxidase assays on both fixed and unfixed tissues.

Due to difficulties in isolating and identifying pneumoviruses, serologic assays have been developed to confirm infection in commercial chickens and turkeys. A number of commercially prepared ELISA kits are available and are used most commonly, but other techniques including virus neutralisation and indirect immunofluorescence have also been used. As with all serologic tests, both acute and convalescent samples should be submitted for analysis.

The sera should be heated to 56°C for 30 min and stored at -20°C if delays in testing are unavoidable. While ELISA that use either subgroup A or B strains as antigens detect antibodies to both of these subgroups, the homologous antigen should be used for the efficient detection of subgroup C.

Differential Diagnosis

Paramyxoviruses (particularly Newcastle disease and paramyxovirus 3), infectious bronchitis virus, and influenza viruses may cause respiratory disease and egg production problems in chickens and turkeys that closely resemble pneumovirus infection. These viruses can be differentiated on the basis of morphology, hemagglutinating and neuraminidase activity, and molecular characteristics. A wide range of bacteria and Mycoplasma spp can cause clinical signs very similar to those of avian pneumovirus. These agents are frequently present as secondary opportunistic pathogens and may mask the presence of the pneumovirus.

Prevention and Treatment

Good management practices can significantly reduce the severity of infection, especially in turkeys; in particular, optimal ventilation, stocking densities, temperature control, litter quality, and biosecurity all have a positive influence on the disease. Some success in reducing disease severity by controlling secondary adventitious bacteria with antibiotics has also been reported. Both live and inactivated vaccines are available for immunisation of chickens and turkeys and are widely used in countries where the disease is endemic. Live vaccines stimulate both local and systemic immunity in the respiratory tract, and cross-protection between subtypes can occur. To produce complete protection in adult birds, oil-adjuvanted inactivated vaccines are administered to birds previously primed with live vaccines.

Myopathies

Minimal Myopathy

A minimal myopathy is seen in otherwise normal meat-type poultry. Affected individuals show no clinical signs and their muscles are grossly normal, but microscopically there is mild myofibre degeneration and fat accumulation between myofibres. Focal or multifocal scattered myofibres are hyalinised and mineralised. More severe examples of this lesion contain individual myofibre necrosis, increased fat, and fibroplasia between fibres. No specific cause has been determined for these minimal changes.

Deep Pectoral Myopathy

Degenerative Myopathy, Green muscle Disease

In this myopathy, there is degeneration, necrosis, and fibrosis of the deep pectoral (supracoracoideus) muscle in heavy meat birds (chickens, turkeys), most often turkey breeder hens. Flock incidence as high as 25% has been reported with few birds developing the disease before 24 wk of age. The major loss is from downgrading or condemnation at processing.

The myopathy may be unilateral or bilateral with the central one-third to two-thirds of the muscle affected. Early, the involved muscle is pale, swollen, and edematous. Later, affected tissue is sharply demarcated from adjacent, viable muscle; eventually, it is encapsulated, resulting in dry, green, necrotic muscle enclosed in a thick fibrous capsule. The defect can be identified by a depression of the breast over affected muscles or by transillumination of carcasses at slaughter.

The deep pectoral muscle functions to elevate the wing. Although well-developed in modern meat birds, it is little used. After episodes of prolonged wing flapping (such as occurs during handling), when lame birds use their wings to assist ambulation, or when the bird is placed on its back, the muscle swells within the dense fascial covering normally surrounding it. This swelling collapses vessels supplying the muscle, leading to ischemia, tissue hypoxia, and muscle necrosis. The lesion can be produced artificially by stimulating the deep pectoral muscle to contract and can be prevented by surgically opening the fascial sheath covering the muscle.

Incidence can be decreased by careful handling of susceptible birds to prevent excessive wing flapping and, as a longterm method, by selective breeding because the condition is heritable. Supplementing rations with selenium, vitamin E, or methionine has not influenced incidence.

Nutritional Myopathy

Nutritional myopathy in poultry, waterfowl, and ostriches is selenium-responsive. Cysteine also may have a beneficial effect in chicks. Lesions have been reported in skeletal, heart, and smooth muscle (gizzard and intestine) of ducks, turkeys, and chickens. Arsenic, zinc, copper, and other metals are antagonistic to selenium, and exposure to these other metals may precipitate outbreaks. Gross lesions, with pale foci or streaking, are similar to those of nutritional myopathies in mammals. Microscopic changes include focal or widespread myofibre swelling, edema, hyalinisation, mineralisation, degeneration, and lysis with infiltration of macrophages and heterophils. Hypercellularity from proliferation of sarcolemmal nuclei may be prominent if regeneration is occurring. Poultry feeds in many parts of the world contain added selenium at 0.1-0.4 ppm to prevent this myopathy.

Toxic Myopathy

Ionophore toxicity causes muscle damage with incoordination, leg weakness, diarrhea, dyspnea, and reduced feed intake and weight. Stunting may also occur. Type I (red or fat-metabolising) fibres are most susceptible. Lesions may also be found in heart and gizzard muscle. Adult birds (chickens, turkeys, ratites) and birds with no previous exposure are more sensitive to ionophore coccidiostats. Gross and histologic changes are similar to those of nutritional myopathy. Toxicity of ionophores is increased if they are used in conjunction with tiamulin, erythromycin, or chloramphenicol.

Cassia (coffee weed) toxicity can produce clinical signs and gross and histologic changes in muscle similar to those seen in ionophore toxicity.

Exertional Myopathy

Myopathy induced by muscle activity develops when local muscle hypoxia or metabolic by products of muscle metabolism (lactic acid), or both, exceed the capacity of the vascular system to remove them. These substances accumulate locally, causing muscle and vascular damage. Mediators of inflammation are activated, resulting in edema and hyperemia.

Exertional myopathy can be caused by isometric muscle activity (e.g., transport, capture, or restraint myopathy as with turkey leg edema syndrome) or kinetic muscle activity (e.g., subgastrocnemial ischemic myopathy of male broiler breeders, polymyopathy of male turkeys). Early gross lesions include pallor with edema or bloodstained transudate. There is swelling, degeneration, necrosis, and mineralisation of muscle fibres, with edema, hemorrhage, and infiltration of heterophils and macrophages.

Mechanically Induced Myopathy

Avascular necrosis caused by pressure in heavy birds that are down because of lameness or leg deformity is seen occasionally and occurs most frequently in the breast muscle. On gross examination, the tissue is firm and pale. Histologic examination reveals swelling, hyalinisation, and necrosis of fibres with edema, heterophils, and macrophages at the periphery.

Rupture of the gastrocnemius tendon is common in meat-type chickens, particularly roasters and breeders, and rare in turkeys. The rupture is due to application of excess weight to tendons that have been previously damaged (most frequently by reoviral, staphylococcal, or toxic tendonitis) with subsequent intra- and peritendonal fibroplasia. This fibroplasia makes the tendon larger but weaker due to replacement of normal strong, dense tendon connective tissue with weak, dense, irregular tissue.

Synechial connections between the tendon and its sheath may also be produced, limiting the tendon's range of motion. Application of normal or excess weight to these previously damaged tendons results in partial or complete tearing or rupture. Rupture of the tendon of one leg puts stress on the other tendon, and bilateral rupture is frequent. Affected birds are lame or "down on their hocks" (creepers).

Hemorrhage from the injury is visible as red, blue, or green discoloration in the tissue above the hock on the back of the leg and results in condemnation of the affected part at processing (red-leg, green-leg). The ruptured tendon can be palpated as a hard mass on the back of the leg above the hock.

Transport Myopathy of Turkeys

Leg Edema Syndrome

Heavy toms are primarily affected, although transport myopathy also develops in hens, especially in flocks in the upper midwest of the USA. About 5% of all flocks are affected, and morbidity within the flock is 2-70%. Transport myopathy occurs sporadically but is most common during fall and early winter. A high incidence has occurred in sequential flocks from the same farm. Incidence is likely to be higher in flocks raised in confinement than in range flocks.

The cause is unknown, but transport myopathy is associated with increased body size and weight, increased transport time to processing plant, cool ambient temperatures, and valgus leg deformities. The pathogenesis is presumed to be due to impaired circulation; the resulting muscle ischemia and acute necrosis lead to edema that tends to be manifested in the large potential space in the subcutis of the medial thigh. Signs are rarely seen on the farm.

Often, only 1 leg is affected. No evidence of external trauma is seen. Skin over edematous subcutaneous tissue is pale, feather follicles are less visible, and the skin slips easily over underlying muscle when moved. Occasionally, there is crepitation. Affected areas are dark when the edematous areas contain blood. Typically, when lesions are cut, the edematous subcutis is a few to several millimetres thick and is amber, occasionally green, or rarely red. Purulent exudate is absent, which distinguishes transport myopathy from cellulitis. If hemorrhage is present, the adductor muscle usually is torn. Removal of affected legs at processing results in carcass downgrading. Microscopically, acute multifocal muscle necrosis is found, primarily in the adductor muscles. Sometimes, subacute or chronic lesions also are seen, which suggest earlier episodes of myopathy. Serum CK increases sharply between farm and processing.

Programs designed to improve leg strength and conformation and to reduce trauma during transportation help reduce the incidence. Supplemental vitamin E also may be useful. If possible, flocks with a high incidence of valgus leg deformities should be marketed early at a processing plant nearby.

Rupture of the Peroneus (Fibularis) Longus Muscle

Rupture of Peroneus (Fibularis) Longus Muscle, Turkey

The origin of the peroneus muscle is on the proximal end of the tibiotarsus and patellar tissue, with attachments to other muscles in that area. In turkeys, the insertion appears to be in 3 places. A small band of tissue from the medial side of the muscle runs to the lateral tibial condyle. The main muscle tendon crosses the lateral side of the hock and joins other tendons that extend the hock and may affect foot and toe movement.

The muscle is thin and wide, covering the anterior and lateral surface of the leg. It has a heavy aponeurosis in which the tendon is embedded. Rupture of the aponeurosis and muscle occurs as a 1-2 cm horizontal wound on the anterior surface of the muscle. It occurs above the middle of the tibiotarsus at the top of the ossifying tendon where the tendon attaches to the muscle. Rupture occurs at 10-14 wk, the age at which turkey leg tendons become ossified, reducing the elasticity of the tissue in that location.

Incidence appears to be increasing. It is most frequent in females and may affect up to 5% of the flock. The separation of the muscle likely occurs slowly, caused by activity such as repeated springing, in turkeys that are becoming heavier and maturing earlier each year. The attachments of this muscle suggest that antagonistic activity is possible. Affected birds are not lame, but the resulting hemorrhage causes a red, blue, or green discoloration under the skin on the anterior of the drumstick ventral to the rupture. The affected portion is trimmed at processing.

Disorders of the Skeletal System

Disorders of the Skeletal System

Skeletal disorders cause lameness from biomechanical dysfunction and result in poor growth, culled birds, increased mortality (caused by starvation and dehydration), and carcass condemnation and downgrading. Production characteristics of modern poultry lines (e.g., body weight in broiler chickens, egg production in laying hens) place high demands on the skeletal system, and inadequacies in nutrition or husbandry will often result in skeletal diseases.

Skeletal disorders may be primarily infectious or noninfectious; both may be seen concurrently within a flock. Before postmortem examination, flocks should be assessed; live, lame birds should be examined, and an opinion as to general flock health, litter quality, and management should be formed. Serum samples may be collected for viral and mycoplasmal serology.

Noninfectious Skeletal Disorders

Rotational (Torsional) and Angular (Valgus/Varus) Deformity

These deformities often are seen as distinct flock problems. Bones all exhibit some degree or combination of lateral, medial, anterior, or posterior bend. They also show some torsion (rotation) about their long axis. The most common abnormalities are valgus deformity of the intertarsal joint and excessive external rotation of the tibiotarsus. Valgus/varus deformity is associated with rapid growth and little exercise. The incidence can be reduced by slowing growth rate at an early age by feed restriction or lighting programs. It may also be due to chondrodystrophy due to B vitamin or trace mineral deficiencies. Rotated tibia has been a major problem in turkeys and a minor problem in Leghorns and guinea fowl. The cause is poorly understood but has been associated with early rickets. Poor mineralisation of the bone, as in rickets, increases the ease of deformation of the bone and therefore the incidence and severity of deformities. Rickets may be associated with nutritional deficiencies, enteric disease, or malabsorption.

Spondylopathies

Vertebral deformities and/or displacements (spondylopathies) are common in thoracic vertabrae, particularly the fifth or free thoracic vertabrae. Spondylolisthesis is the most common deformity, but incidence is low in most flocks of broiler chickens. It causes posterior paralysis due to spinal cord compression.

Dyschondroplasia

Dyschondroplastic lesions are masses of avascular cartilage extending from the growth plate into the metaphysis and are attributed to the failure of chondrocytes to differentiate. This results in a focal thickening of the growth plate in the proximal tibiotarsus (tibial dyschondroplasia) or sometimes the proximal tarsometatarsus.

The lesion in the proximal tibiotarsus is often associated with anterior bowing of the tibiotarsus and sometimes fractures below the plug of cartilage. Factors shown to influence the incidence and severity of dyschondroplasia include genetic selection, calcium:phosphorus ratios in feed, metabolic acidosis through excess chloride in feed, acid/base balance, and mycotoxins. In a flock of modern broilers, the cause may be marginal inadequacies in dietary calcium or a calcium:phosphorus imbalance.

Rickets

Rickets develops in growing birds due to deficiency of calcium or phosphorus (*Calcium and Phosphorus Imbalances*) or insufficient vitamin D (*Vitamin D3 Deficiency*). Malabsorption can also cause a mineral deficiency. In rickets, a failure of bone mineralisation leads to flexibility of long bones. Bone ashing and estimates of calcium and phosphorus content combined with bone pathology are useful diagnostic tools. Bacterial infections are common in bones with rickets.

Plantar Pododermatitis

Ulceration of the metatarsal and digital footpads is a common cause of lameness in meat-type poultry. Wet or poor quality litter is the common cause, although a biotin deficiency will cause plantar pododermatitis even when litter quality is good. Ulcerated footpads may become secondarily infected and caked with litter.

Osteopenia (Osteoporosis and Osteomalacia)

Osteopenia is a consequence of osteoporosis, a deficiency in the quantity of fully mineralised, structural bone. Cage layer fatigue describes a syndrome in which laying hens become paralysed in their cages. The bones of the birds are osteopenic. The sternum is often deformed, and fractures causes infolding of the ribs at the junctions of the sternal and vertebral portions. Fractures can also occur in the long bones and vertebrae.

The medullary bone is osteomalacic. The syndrome is due in part to a lack of exercise and high egg production, but severe problems are associated with inadequate calcium, phosphorus, or vitamin D. Calcium requirements during growth and before and during lay vary markedly. Sources of calcium that enable the slow release of mineral, such as oyster shell, appear to give the best results.

Amyloidosis

Extensive amyloid arthropathy is primarily caused by Enterococcus faecalis, but not by all isolates. Clinical cases are seen only occasionally and are most frequently seen in the hock joint of a few replacement pullets or broiler breeders. Cases may be attributed to the contamination of a previously sterile vaccine diluent with E faecalis during administration (e.g., Marek's vaccine in day-old chicks).

Infectious Skeletal Disorders

Control

Coagulase-positive staphylococci are frequently responsible for bacterial infections in the bones and joints of broiler chickens. Mycoplasma synoviae (*Mycoplasma synoviae Infection)* may also play a role in infectious bone disorders and can be monitored serologically.

In broilers, bacterial infections are most common in the proximal femur and proximal tibiotarsus when the birds are >22 days of age. In the proximal femur, the condition is also referred to as femoral head necrosis. Recent reports indicate this is the most common cause of lameness in broilers. The etiology appears dependent on vertically transmitted staphylococci in combination with a challenge by immunosuppressive viruses (e.g., infectious bursal disease, *Infectious Bursal Disease: Introduction*).

Floor eggs have been shown to be common carriers of staphylococci, so their use should be minimal. A high standard of hatchery hygiene can reduce this risk. Formaldehyde fumigation within the hatchers is also likely to help. In addition, hatchery fluff samples can be examined to monitor for contamination with staphylococci. Staphylococcal infections in joints and tendons are also seen in breeders. Outbreaks are likely to be due to management practices or other diseases causing stress and/or joint and tendon trauma (e.g., competition over feed space, heavy coccidiosis challenge). Insufficient lighting in the rear of cages appears to predispose to an increase in bacterial tenosynovitis.

Escherichia coli is often responsible for flock outbreaks of arthritis and osteomyelitis in broiler chickens and turkeys. These outbreaks may be associated with respiratory disease. Pasteurella multocida has been isolated from arthritic joints in broiler breeders following use of live vaccines. Other sporadic causes of arthritis in poultry include Salmonella spp and Streptobacillus moniliformis. Viral arthritis due to a reovirus has been reported as a significant cause of lameness in some parts of the world. The virus is egg transmitted. Vaccines against the condition have been developed.

Control

Bacterial bone and joint infections often show a poor response to antibiotic treatment. Antibiotics may be used to control the bacteremia contributing to new cases and to modify the bacterial flora within a flock. When individual birds are of high value, injections of long-acting antibiotics may improve some less severe cases. Control requires minimising sources of infection and stock susceptibility.

West Nile Virus Infection in Poultry

West Nile virus (WNV), a flavivirus related to the St. Louis encephalitis/Japanese encephalitis complex, was first isolated from the blood of a febrile Ugandan woman in 1937. The virus was first described as the cause of a West Nile fever epidemic in humans in Israel in 1951; in a later outbreak, severe meningoencephalitis was seen in elderly patients.

The role of mosquitos in viral transmission was clearly delineated in a series of field studies in Egypt in the 1950s. Wild birds were identified as the reservoir of the virus around the same time. Cases of West Nile fever in horses were reported several years later. WNV was first associated with disease in domestic avian species in 1997, when flocks of young geese in Israel were affected with a neuroparalytic disease. In August 1999, the disease appeared for the first time in the Western Hemisphere when wild and zoo birds, horses, and humans died in the northeast USA, notably in the New York City area.

Etiology and Epidemiology

WNV is considered to be endemic in many countries of Africa, Asia, southern Europe, and North America. Epidemics appear in the human population at infrequent intervals in some of these countries, and there is evidence for viral transmission between Africa and Europe by migrating birds. Most outbreaks have occurred from mid-July through October when cold nights reduce mosquito vector activity, notably Culex spp. Geese are the only known natural hosts of WNV among domestic avian species. Most of the flocks affected in the Israel outbreaks were 5-9 wk old, but goslings as young as 3 wk and as old as 11 wk were also affected. Adult breeding flocks were clinically unaffected but virus neutralising antibodies were found. Mortality of young Muscovy ducks but not young chickens or turkey poults was induced experimentally with a WNV isolate.

Transmission

The principal route of viral transmission is by the bite of a mosquito (primarily Culex spp). In the USA during 1999 and 2000, most of the viral isolates were made from C pipiens and C restvans. In Africa and the Middle East, the usual vector is C univittatus, and in Europe, C pipiens and C modestus. WNV has also been isolated from at least 10 tick species.

Clinical Findings

Affected geese show various degrees of neurologic involvement ranging from recumbency to leg and wing paralysis. Affected birds are either reluctant or unable to move when disturbed. Signs of incoordination are pronounced and some birds flip over while attempting to stand. Naturally affected geese show torticollis and opisthotonos. Mortality rates of 20-60% have been reported, probably due to horizontal spread of the virus.

Lesions

Pathologic changes include pallor of the myocardium and occasionally of the kidneys, splenomegaly, and hepatomegaly. The meningeal blood vessels are injected. Microscopic brain lesions consist of lymphocytic perivascular infiltration and neuronal degeneration. Small necrotic foci are present in the myocardium but lymphocytic infiltration is minimal.

Diagnosis

The tissues of choice for isolating virus from paralytic or dead birds are the brain, spleen, and kidneys. Homogenates are inoculated into the brain of newborn mice, embryonated eggs by the yolk sac route, or Vero and mosquito cell line cultures. Reverse transcriptase PCR with RNA extracted from either brain material or cell culture supernatant can also be performed. TakMan technology for rapid molecular diagnosis of field-collected mosquitos and avian tissues has been introduced recently. Immunohistochemistry can be used on formalin-fixed paraffin-embedded tissues, in particular brain and kidney, to visualise viral antigens in infected birds. Several forms of ELISA have also been developed for flaviviruses.

Neurologic signs in young geese must be distinguished from those caused by Riemerella (Pasteurella) infections, especially R anatipestifer. Other bacteria include Streptococcus gallolyticus, and Erysipelothrix, Listeria, and Salmonella spp. Neurotropic viruses include Newcastle disease, which is rare in geese, and avian influenza. Ionophore intoxication can induce paralytic signs. Aspergillus also causes brain lesions and caseous nodules in the lungs.

Prevention and Control

Mosquito control is a mandatory component of any arboviral disease control program. Unfortunately, this is difficult to implement in a rural environment because of the distances that mosquitos can fly or be carried by prevailing winds. Standing water and similar insect breeding sites in the vicinity of densely populated avian farms should be treated with larvicides. Poultry houses should be constructed to be insect free. Because many arboviral diseases are zoonoses, much can be achieved by cooperation with human disease surveillance agencies.

Control of WNV in geese is confined to vaccinating young flocks at risk, especially those raised during July through November when Culex spp are most numerous. Because of confounding factors such as possible horizontal transmission of virus, all birds in the flock should be vaccinated. Due to the age-related susceptibility to the virus, goslings should be immunised as young as possible, preferably at 3 wk old. Currently, WNV vaccines are not available commercially, although several types have been developed.

Laboratory trials have been performed with a formaldehyde-inactivated suckling mouse brain-derived product. Over 75% of geese vaccinated with a single dose of vaccine at 3 wk of age were protected, and 94% protection was achieved with 2 doses spaced 2 wk apart.

The duration of immunity was estimated to last until 12 wk of age. Inactivated vaccines prepared from chick embryos or Vero cells are not protective because of their low antigenic mass. A single dose of a live attenuated vaccine derived from high mosquito cell passage induced immunity to intracerebral challenge in young geese. Mosquito feeding experiments and back passage reversions to virulence studies have yet to be completed.

Ectoparasites

Bedbugs

Cimex lectularius is a common bloodsucking parasite in temperate and subtropical climates that attacks poultry, humans, and most other mammals. It is rare in modern laying operations, but breeding houses and pigeon lofts may become heavily infested. The life cycle may be completed in 4-6 wk or extend much longer because nymphs can withstand fasting for ~70 days, and adults for up to 12 mo. Feeding usually occurs at night. Bedbugs become engorged within 10 min, then hide in cracks and crevices. If attacked by large numbers of bedbugs, birds may become anemic. Bites are usually followed by swelling and itching due to the injection of saliva into the wound.

Control is best accomplished by thoroughly cleaning the houses, reducing hiding places for the bedbugs, and high-pressure spraying of the houses as for control of fowl ticks.

Fleas

The sticktight flea, Echidnophaga gallinacea, is unique among poultry fleas in that the adults become sessile parasites and usually remain attached to the skin of the head for days or weeks. The adult females forcibly eject their eggs so that they reach surrounding litter. The larvae develop best in sandy, well-drained litter. Hosts of the adult flea include chickens, turkeys, pigeons, pheasants, quail, humans, and many other mammals. Irritation and blood loss may cause anaemia and death, particularly in young birds.

The Western hen flea, Ceratophyllus niger, seems to be confined to the Pacific coast area of the USA. This flea actually breeds in droppings and feeds on birds only occasionally. The European chick flea, C gallinae, is widespread in the USA. It breeds in nests and litter and is on the birds only to feed. It attacks many other birds besides chickens. The most important control measures are removing infested litter and dusting the litter surface with carbaryl, coumaphos, or malathion to kill immature fleas. Insect growth regulators such as methoprene are also effective. Sticktight fleas can be controlled by topical application of pyrethrin.

Mites

The most economically important of the many external parasites of poultry are mites of the families Dermanyssidae (chicken mite, northern fowl mite, and tropical fowl mite) and Trombiculidae (turkey chigger).

Chicken Mite

Red mite, Roost Mite, Poultry Mite

Dermanyssus gallinae infests chickens, turkeys, pigeons, canaries, and various wild birds. While rare in modern commercial cage-layer operations, it is found in breeder and small farm flocks. Chicken mites are nocturnal feeders that hide during the day under manure, on roosts, and in cracks and crevices of the chicken house, where they deposit eggs. Populations develop rapidly during the warmer months and more slowly in cold weather; the life cycle may be completed in only 1 wk. A house may remain infested for 6 mo after birds are removed.

Dermanyssus Gallinae, Skin Lesions

Dermanyssus Gallinae

Transmission of the chicken mite, as well as the northern fowl mite and the tropical fowl mite, is by mite dispersion or by contact with infested birds, animals, or inanimate objects. In the integrated poultry industry, mites are dispersed most frequently on inanimate objects such as egg flats, crates, or coops, or by personnel going from house to house or farm to farm.

Heavy infestations of either chicken mites or northern fowl mites decrease reproductive potential in males, egg production in females, and weight gain in young birds. Chicken mites may be found in the chicken houses during the day, particularly in cracks or where roost poles touch supports, or on birds at night. Northern fowl mites are found on eggs or by parting feathers in the vent area.

Obtaining mite-free birds and using good sanitation practices are important to prevent a buildup of mite populations. Once poultry have been infested with northern fowl mites, control may be achieved by spraying or dusting the birds and litter with carbaryl, coumaphos, malathion, stirofos, or a pyrethroid compound in areas where the parasites have not developed resistance to these chemicals. Miticide spray treatments must be applied with sufficient force to penetrate the feathers in the vent area. Nicotine sulfate is an effective fumigant for mites but is particularly hazardous. Pyrethrins and piperonyl butoxide are initially active but have poor residual killing power.

For control of chicken mites, in addition to treating the birds, the inside of the house and all hiding places for the mite (such as roosts, behind nest boxes, and cracks and crevices) must be treated thoroughly using a high-pressure sprayer. Dimethoate and fenthion may be used as residual house sprays when poultry are not present. Systemic control with ivermectin (1.8-5.4 mg/kg) or moxidectin (8 mg/kg) is effective for short periods, but the high dosages are expensive, close to toxic levels, and require repeated use.

Feather Mite

Most feather mites belong to the families Analgidae (Analgesidae), Pterolichidae, and Proctophyllodidae and are rare on modern poultry ranches. They do little economic damage but may reduce egg production via malnutrition, feather loss, and dermatitis. Affected birds should be dusted with pyrethrin or carbaryl powder.

Scaly Leg Mite

Knemidocoptes Mutans, Skin Lesions

Knemidocoptes mutans is a small, spherical, sarcoptic mite that usually tunnels into the tissue under the scales of the legs. It is rare in modern poultry facilities, but when found, it is usually on older birds on which the irritation and exudation cause the legs to become thickened, encrusted, and unsightly. This mite may occasionally attack the comb and wattles. The entire life cycle is in the skin; transmission is by contact. For control, affected birds should be culled or isolated, and houses cleaned and sprayed frequently as recommended for the chicken mite. Individual birds should be treated with oral or topical ivermectin.

Subcutaneous Mite

Laminosioptes cysticola is a small parasite that is most often diagnosed by observing white to yellowish caseocalcareous nodules ~1-3 mm in diametre in the subcutis. Careful examination of the skin and subcutis of birds under a dissecting microscope frequently reveals the mites. Destroying the bird has been the best control for this parasite, but ivermectin may be effective.

Turkey Chigger

The larvae of Neoschongastia americana are parasitic on numerous birds. Across the southern USA, they are the major pest of turkeys ranged on heavy clay soils in the summer.

The chiggers feed in groups of up to 100 mites/lesion for 8-15 days. Turkeys may have 25-30 lesions each. One lesion, 3 mm in diametre, may cause significant downgrading at market time. To prevent downgrading, turkeys must be protected for e"4 wk before marketing.

Sprays or dusts of malathion or chlorpyrifos on turkey ranges control chiggers. A preventive measure now used in many turkey-growing areas includes a shift from range to confinement rearing, or use of sheds to provide shade.

Disposition and Fate of Drugs: Overview

Once a drug has been administered by any route other than IV, it must be absorbed into the bloodstream from the site of administration. The drug then is distributed into various body fluids and tissues to attain an effective, yet safe, concentration for a sufficient period of time at the site of action. Subsequently, the drug is inactivated or eliminated from the body, generally by metabolism (usually lipid-soluble drugs) and excretion (mainly renal and biliary routes). The effectiveness of these processes with respect to time (pharmacokinetics) varies with the particular drug and species of animal. It is equally influenced by disease and the effects of concurrently administered agents (drug interactions).

Drug Absorption

Passage of Drugs Across Cellular Membranes:

Regardless of the route of administration, a drug usually must cross a number of membranes before it reaches its site of action. Membrane barriers may be composed of several layers of cells (e.g., skin, vagina, cornea, placenta) or a single layer of cells (e.g., enterocytes, renal tubular epithelial cells), or they may consist only of a boundary <1 cell in thickness (e.g., hepatic sinusoids, single cell membrane, mitochondrion, nucleus).

Drugs and other molecules can cross cellular membranes by several processes. Passive transfer or simple diffusion is the most important for xenobiotics, although specialised transport systems are used for a limited number of therapeutic agents.

In simple diffusion, movement of the drug is due to and directly related to its concentration gradient across the membrane. In the case of lipid diffusion, lipid-soluble substances dissolve in the lipid phase of the membrane and diffuse down their concentration gradients into the aqueous phase on the other side of the barrier. Thus, the ability of a compound to cross a membrane by simple lipid diffusion is a function of its degree of lipid solubility (lipid-to-water partition coefficient). The molecular mass of the drug, the thickness of the membrane(s), and the surface area available also influence the rate of diffusion.

Many agents of pharmacologic interest are weak organic electrolytes. At physiologic pH, these weak acids or bases may be present partly in the ionised (dissociated) and partly in the nonionised (undissociated) form. The ratio between the respective forms depends on the drug's dissociation constant (pK_a) and the pH of the solution in which it is dissolved.

The nonionised fraction may penetrate biologic membranes by lipid diffusion and become distributed across the membrane according to the degree of ionisation on each side of the membrane and the extent to which the drug is bound to proteins or other macromolecules in the solutions bathing either side of the membrane.

Membranes are more permeable to the undissociated molecule than to the ionised form, simply because the nonionised form is much more lipid soluble. Although a compound may be nonionised, it also may be so poorly soluble in lipids that it penetrates biologic membranes only to a limited extent. A degree of aqueous solubility is also necessary for a drug to be in solution in the body fluids on either side of a cellular membrane.

It is supposed that aqueous pores exist in lipoproteinaceous biologic membranes. Lipid-insoluble compounds can easily diffuse through these pores, as well as directly through the membrane, at rates that depend on their molecular masses and concentration gradients. However, with ions or other polar compounds, the speed of transfer is determined by both the charge and molecular dimensions of the drug. When a hydrostatic or osmotic pressure difference exists across a membrane, water flows through the aqueous pores; this bulk fluid movement carries or "drags" solute molecules through the pores in the moving stream, provided that the solute molecules are smaller than the aqueous channels.

Several specialised transfer processes account for the passage of certain organic ions and other large lipid-insoluble substances across biologic membranes. Active transport, facilitated diffusion, and exchange diffusion are 3 distinct types of carrier-mediated systems used for moving specific substances across cellular membranes. The highly selective carrier-mediated systems are principally used for transporting nutrients and natural substrates across biologic membranes.

Pinocytosis is an important transport process in mammalian cells, particularly intestinal epithelial cells and renal tubular cells. Drugs that exist in solution as molecular aggregates, have large molecular masses themselves, or are bound to macromolecules may be transferred across membranes by pinocytosis.

Drug Absorption from the GI Tract

Although the basic principles governing the absorption of drugs from the GI tract are understood, many confounding factors may play a role in modifying the process, and erratic responses may result. Some of the more important factors to be considered include the following:

1) molecular size and shape of the drug and its concentration,
2) degree of ionisation at specific pH values (depends on pK_a of the drug),
3) lipid solubility of the neutral or nonionised form of the drug,
4) chemical or physical interactions with coadministered drug preparations or even food constituents,
5) the pharmaceutical preparation and characteristics of the dosage form (especially the disintegration and dissolution rates of solid dosage forms),
6) morphologic and functional differences of the GI tract among the various animal species,
7) gastric motility, secretion, and the rate of gastric emptying,
8) intestinal motility and secretions as well as the intestinal transit time,
9) fluid volume within the GI tract,
10) osmolality of intestinal content,
11) intestinal blood and lymph flow,
12) disruption of the structural and functional integrity of the gastric and intestinal epithelium, and
13) drug biotransformation within the intestinal lumen by microflora, or within the mucosa by host enzyme systems.

Bioavailability

This term is used to define the rate and extent to which a drug administered in a particular dosage form enters the systemic circulation intact. All of the considerations outlined above, as well as the particular product used, can influence bioavailability. Biotransformation by intestinal epithelial cells, and particularly by liver cells, can substantially reduce the amount of unchanged drug that enters the systemic circulation after administration PO. This is known as the "first-pass" effect and is significant for a number of drugs.

Drug Absorption from Topical Administration

Drugs may be absorbed through the skin after topical application; however, the stratum corneum presents an effective barrier to movement of most drugs. The intact skin allows the passage of small lipophilic substances but efficiently retards the diffusion of water-soluble molecules in most cases. Lipid-insoluble drugs generally penetrate the skin slowly in comparison with their rates of absorption through other body membranes. Absorption of drugs through the skin may be enhanced by inunction or more rarely by iontophoresis if the compound is ionised. Certain solvents (e.g., dimethyl sulfoxide [DMSO]) may facilitate the penetration of drugs through the skin. Damaged, inflamed, or hyperemic skin allows many drugs to penetrate the dermal barrier much more readily. The same principles that govern the absorption of drugs through the skin also apply to the application of topical preparations on epithelial surfaces.

Drug Absorption from Tracheobronchial Surfaces and Alveoli

Because volatile and gaseous anesthetics have relatively high lipid-to-water partition coefficients and generally are rather small molecules, they diffuse practically instantaneously into the blood in the alveolar capillaries. Particles contained in aerosols can be deposited, depending on the size of the droplets, on the mucosal surface of the bronchi or bronchioles, or even in the alveoli. Most drugs are usually absorbed quite rapidly from these sites according to the principles discussed above.

Drug Absorption from Parenteral Delivery Sites

After a drug has penetrated the skin, GI epithelium, or other absorbing surface, or has been deposited by injection into a body tissue, it comes into the immediate vicinity of capillaries. Solutes traverse the capillary wall by a combination of 2 processes: diffusion and filtration. Diffusion is the predominant mode of transfer for lipid-soluble molecules, small lipid-insoluble molecules, and ions. All drugs, whether lipid-soluble or not, cross the capillary wall at rates that are extremely rapid compared with their rates across other body membranes. In fact, the movement of most drug molecules in various tissues is limited only by the rate of blood flow rather than by the capillary wall. However, some endothelial cells, such as the blood-brain barrier, have much tighter intercellular junctions than others and, therefore, restrict drug movement more significantly.

Aqueous solutions of drugs are usually absorbed from an IM injection site within 10-30 min, provided blood flow is unimpaired. Faster or slower absorption is possible, depending on the concentration and lipid solubility of the drug, vascularity of the site (there are differences between various muscle groups), the volume of injection, the osmolality of the solution, and other pharmaceutical factors.

Substances with molecular weights >20,000 daltons are principally taken up into the lymphatics. Absorption of drugs from subcutaneous tissues is influenced by the same factors that determine the rate of absorption from IM sites. Some drugs are absorbed as rapidly from subcutaneous tissues as from muscle, although absorption from injection sites in subcutaneous fat is always significantly delayed.

Increasing blood supply to the injection site by heating, massage, or exercise hastens the rate of dissemination and absorption. Spreading and absorption of a large fluid volume that has been injected SC may be facilitated by including hyaluronidase in the solution. The rate of absorption of an injected drug may be prolonged in a number of ways, including immobilisation of the site, local cooling, a tourniquet, incorporation of a vasoconstrictor, an oil base, and implant pellets and other insoluble "depot" preparations. Among these depot preparations are drugs that are converted to less soluble salts (e.g., procaine and benzathine penicillin) or less soluble complexes (e.g., protamine zinc insulin), or that are administered as insoluble microcrystalline suspensions (e.g., methylprednisolone acetate).

Drug Distribution

After absorption into the bloodstream, drugs become disseminated to all parts of the body. Compounds that permeate freely through cell membranes become distributed, in time, throughout the body water, both extracellular and intracellular. Substances that pass readily through and between capillary endothelial cells, but do not penetrate other cell membranes, are distributed into the extracellular fluid space. Occasionally, the drug molecule may be so large (>65,000 daltons) or so highly bound to plasma proteins that it remains in the intravascular space after IV administration. Drugs may also undergo redistribution in the body after initial high levels are achieved in tissues that have a rich vascular supply, e.g., the brain. As the plasma concentration falls, the drug readily diffuses back into the circulation to be quickly redistributed to other tissues with high blood-flow rates, such as the muscles; then, over time, the drug also becomes deposited in lipid-rich tissues with poor blood supplies, such as the fat depots. Most drugs are not distributed equally throughout the body but tend to accumulate in certain specific tissues or fluids.

The general principles that govern the passage and distribution of drugs across cellular membranes are applicable. Basic drugs tend to accumulate in tissues and fluids with pH values lower than the pK_a of the drug; conversely, acidic drugs concentrate in regions of higher pH, provided that the free drug is sufficiently lipid soluble to be able to penetrate the membranes that separate the compartments. Even small differences in pH across boundary membranes, such as those that exist between CSF (pH 7.3) and plasma (pH 7.4), milk (pH 6.5-6.8) and plasma, renal tubular fluid (pH 5.0-8.0) and plasma, and inflamed tissue (pH 6.0-7.0) and healthy tissue (pH 7.0-7.4), can lead to unequal distribution of drugs with pK_a values close to those of the pH of the fluid. Only freely diffusible and unbound drug molecules are able to pass from one compartment to another. Binding to macromolecules such as protein components of cells or fluids, dissolution in adipose tissue, formation of nondiffusible complexes in tissues such as bone, incorporation into specific storage granules, or binding to selective sites in tissues all impede movement of drugs in the body and account for differences in the cellular and organ distribution of particular drugs. Therapeutic agents may also be transported by carrier-mediated systems across certain cellular membranes, which leads to higher concentrations on one side than the other. Examples of such nonspecific transport mechanisms are found in renal tubular epithelial cells, hepatocytes, and the choroid plexus.

Only the unbound or free fraction of a drug can diffuse out of capillaries into tissues. The most important binding of drugs in circulation is to plasma albumin, although the globulins and, especially, á-1 acid glycoprotein (for bases) may also play a significant role. A drug may become bound to plasma proteins to a greater or lesser degree, depending on a number of factors, e.g., plasma pH, concentration of plasma proteins, concentration of the drug, the presence of another agent with a greater affinity for the limited number of binding sites, and the presence of acute-phase proteins during active inflammatory conditions. The degree of plasma-protein binding and the affinity of a drug for the nonspecific protein-binding sites is of great clinical significance in some instances and much less so in others.

For example, a potentially toxic compound (such as dicumarol) may be 98% bound, but if for any reason it becomes only 96% bound, then the concentration of the free active drug that becomes available in the plasma is doubled, with potentially harmful consequences. The concentration of a drug administered in overdose may exceed the binding capacity of the plasma protein and lead to an excess of free drug, which can diffuse into various target tissues and produce exaggerated effects. Of equal importance is the readiness with which drugs dissociate from plasma proteins. Those that are more tightly bound tend to have much longer elimination half-lives because they are released gradually from the plasma protein reservoir. The long-acting sulfonamides are good examples of this phenomenon. Most unbound drugs distribute easily to extracellular fluid. All membranes are transversed only by the more lipid-soluble drugs. During distribution and elimination from the body, a drug may or may not penetrate certain "physiologic" (e.g., blood-brain, placental, and mammary) barriers. A drug may gain access to the CNS by 2 distinct routes—the capillary circulation and the CSF. Drugs penetrate into the cortex more rapidly than into white matter, probably because of the greater delivery rate of drug via the bloodstream to the tissue. The pharmacologic factors and consequences of the diverse rates of entry of different drugs into the CNS include the following: 1) water-soluble ionised drugs will not enter the CNS; 2) low ionisation, low plasma-protein binding, and a fairly high lipid-water partition coefficient confer ready penetration; 3) direct injections into the CSF often produce unexpected effects; and 4) meningoencephalitis can substantially alter the permeability of the blood-brain barrier.

The placental barrier should be considered when selecting an agent to treat a pregnant animal. The potential teratogenicity of any drug needs to be known before its administration; if it is to be used during late gestation, its effects on the fetus and on the process of parturition should be considered. Nutrients, such as glucose, amino acids, minerals, and even some vitamins, are actively transported across the placenta. The passage of drugs across the placenta is largely by lipid diffusion, and the factors discussed above play a role. The distribution of drugs within the fetus follows essentially the same pattern as in the adult, with some differences with respect to the volumes of drug distribution, plasma-protein binding, blood circulation, and greater permeability of interceding membranous barriers.

The mammary gland epithelium, like other biologic membranes, acts as a lipid barrier, and many drugs readily diffuse from the plasma into milk. The pH of milk varies somewhat, but in goats and cows it is generally 6.5-6.8 if mastitis is not present. Weak bases tend to accumulate in milk because the fraction of ionised, nondiffusible drug is higher. The opposite is true for acidic drugs. Agents delivered by intramammary infusion can diffuse into plasma to a greater or lesser degree by the same processes noted earlier.

Chapter 6

Commercial Poultry Project Cycle

Organised groups of poor women are provided credit from the Proshika Revolving Loan Fund.

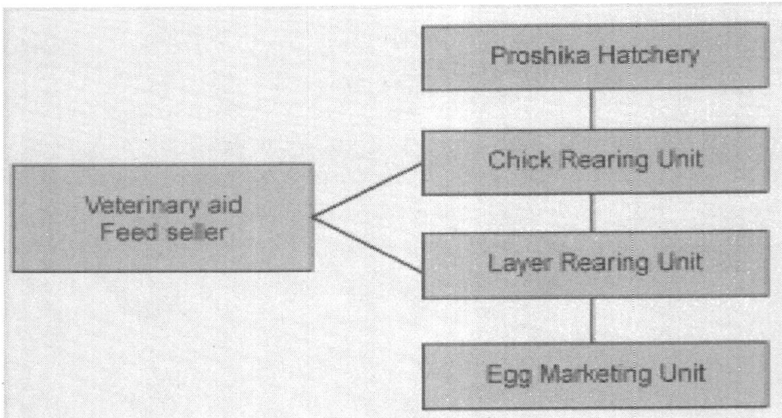

Figure: *Units used in implementation of the poultry project.*

One study found that a group member could earn a net profit of Taka 1,000 per month through a Chick Rearing Unit besides doing all her household work. From the Layer Rearing Project a woman can earn a net profit of Taka 800 to 900 per month.

An impact assessment survey was conducted during the period December 1994 - February 1995. The Impact Monitoring and Evaluation Cell (IMEC) of Proshika and the Horizon Pacific International prepared the Report jointly. The report showed that in female-headed households included in the project earnings on average are 68.1% higher than in control households. Savings are 409% higher.

Proshika Collaboration Projects and its Impact

Proshika has been operating three collaborative projects. A large collaboration project is on the "Participatory Livestock Development Project" (PLDP) in 28 thanas. This is recently undertaken by Proshika in collaboration with Directorate of Livestock Service, financed by Asian Development Bank (ADB) & DANIDA. Proshika also has operated collaborative projects on " Poultry for Nutrition" (PNP) in 17 thanas with the Ministry of Fisheries & Livestock, financed by the World Bank.

Proshika has recently completed a collaborative, three year project entitled "Smallholder Livestock Development Project" (SLDP) with the Directorate of Livestock services (DLS), Govt. of Bangladesh and financed by DANIDA & the International Fund for Agricultural Development (IFAD).

The target beneficiaries of all the aforementioned projects are poor and mostly women. The strategy is to provide microfinance and technical services through NGOs for livestock enterprises suitable for the poor and to develop the capabilities of DLS, NGOs and rural communities to plan and manage livestock development activities and in the process alleviate poverty and improve economic and nutritional status of the rural poor through self-employment. Saleque has detailed the organisational setup in these proceedings.

Impact

Of these three projects, SLDP has been completed. The project was originally designed to undertake a range of economic activities in the context of small-scale livestock development, but ended up concentrating mainly on activities related to poultry development. However, the integration of other species of livestock with poultry enterprises has already been introduced in the PLDP & PNP projects. The project is based model develop by DLS and BRAC.

An impact study conducted by Bangladesh Livestock Research Institute (BLRI) (Alam, 1997) in 5 districts of Bangladesh evaluated the impact of interventions made by the (SLDP) on the socioeconomic conditions of the poor people. The study found that the average size of the loan received per beneficiaries was Taka 2588 with a range from Taka 1,000 to Taka 10,000, depending on the type enter-prise. The loans were repaid by weekly or monthly instalments the repayment rate was 99 per cent. The total net income per household was Taka 455.3 and the aver-age net income per household from SLDP activities was Taka 102.1 per week. The average weekly income of beneficiary households has increased by 31 per cent after membership.

With the increase in income, the households made substantial progress in savings. The total cumulative savings per beneficiary after membership was Taka 1475.7, which was made up of Taka 568 form group savings and Taka 908 from own savings.

At the same time, the consumption of all food items and investment in assets has significantly increased. The project has ensured empowerment of women in the study areas and increased their participation in household decision-making. The generation of income and employment from the SLDP activities has enhanced the status of women in the family.

Constraints and its Probable Solution

The constraints for implementation of poultry projects at the farmers level are as follows-

- Lack of quality feed supply
- Lack of vaccines especially gumboro, Infectious Bursal Disease (IBD) and Marek's disease
- Low price of dressed broilers and eggs.

In order to overcome these constraints, Proshika has developed 2-3 feed seller projects in every Thana. Proshika trains group members to undertake these feed seller projects. They receive credit from Proshika and keep concentrate ingredients for the poultry rearers. Feed formulation, quality and the Proshika livestock technical workers ensure good storage conditions. As vaccines against gumboro, Marek and IBD are not produced locally; Proshika is importing all these vaccines from abroad. To ensure good prices of eggs and broilers by the farmers, Proshika has set up a marketing venture in Dhaka where eggs and broilers are collected from the farmers by group members and sold.

Recommendations

In order to increase the poultry production by the poor people, the following recommendations should be implemented:

- The Government Department of Livestock Services should produce all types of vaccines required for commercial breeds and increase the local vaccine production.
- The Government should subsidize the cost of importing yellow maize, soy-bean meal, and concentrate from abroad.
- A well-organised marketing system should be established to give the farmers a better price for their livestock products.

- The National media (Radio, TV) should arrange a campaign for successful poultry projects.
- Regional workshops, seminars, exhibitions should be organised at the inter-national level for sharing the experiences of farmers about successful project.
- Regional Livestock Training Institute can be setup to increase the capacity of NGO and private sectors.
- Co-ordination amongst Government, NGOs and donors should be strengthened.
- Poultry farm owners should be encouraged to set up more hatcheries to ensure the chicks supply as per farmers' demand.
- Govt. policy for importing input (feed, medicine, vaccines etc.) should be favourable and Value added Tax (VAT) should be withdrawn.

Mass poverty in Bangladesh stems from structural deprivation and is caused by the interplay of several factors.

Proshika has tried to address these factors in its different development strategies. Poultry development is one of the potential areas under Proshika's livestock development programme, which has a potential for capturing the inequitable distribution of income and employment in rural areas.

In Proshika's experiences, women are able to operate and manage impressive technical enterprises like broiler, layer, duck farms etc. efficiently with a high return on the investment.

On the other hand, poultry production under SLDP, PNP and PLDP are useful to improve the native backyard poultry under scavenging and semi-intensive systems, where women traditionally play the most important role.

International Replication of the Grameen Bank

Lessons of Experience for other Development Programs for Poor Women

Is there a general theory and practice of international replication of successful poverty-reduction programs? Almost certainly not, as there appears to be as much art in the process as there is science. Nevertheless, some guidelines can be drawn from the international replication of the Grameen Bank (GB) methodology for poverty-reduction and the empowerment of poor women in Asia, that should be of value in the process of replicating other development programs for poor women in the same region, like the multi-institutional Poultry Production and Health Program (PPHP) located at the Danish Royal Veterinary and Agricultural University.

To discover these, we need to take a close look at the process of replication of the Grameen Bank.

International Replication of the Grameen Bank

Replications and adaptations of the Grameen Bank methodology have sprung-up, like mushrooms, all over the world. Such is the power of the example set by the Grameen Bank and of young people everywhere who want to believe that a world without absolute poverty is possible and who are willing to work for it.

Many of these efforts are successful on a small scale, in that poor women are using the financial services offered to create self-employment that increases household income and reduces their poverty. At the same time they are repaying their small loans faithfully and building-up their tiny savings.

Such is the power of the desire of poor women everywhere for a better life for their children. In total, however, the number of poor households in the world being reached (by all types of micro-finance for the poor) remains relatively small, certainly less than ten million, especially in relation to the more than two hundred million households still living in absolute poverty today. Most of the households being reached are in Bangladesh. CASHPOR Inc., the network of GB replications and adaptations in Asia, has been promoting the international replication of GB in the Asian region outside of Bangladesh.

Cashpor Inc

From six programs in four countries reaching together about 20,000 poor house-holds, when the network was formed in September 1991, it has grown to 22 pro-grams in 8 countries together reaching more than 250,000 poor women and their households. Details can be seen in the CASHPOR Member Update that appears in every issue of *Credit for the Poor,* the quarterly newsletter of CASHPOR Inc. Almost 35 million US dollars is outstanding in the hands of more than 200,000 poor women, and together the 250,000 served have saved nearly 5 million US dollars. Repayment rates in most programs are near perfect, with only just over 1% of the total loans outstanding at risk.

Cashpor promotes the international replication and adaptation of GB in two main ways:

1) by providing training and technical services to increase the institutional capacity of its member institutions, all of whom are hands on replicators and adopters (GBRs); and
2) by establishing new replications and adaptations where there are large numbers of poor households not yet served by a GBR.

An example of the latter is our current experimental replication and adaptation, CASKPOR Financial and Technical Services Pvt. Ltd. (CFTS), in Mirzapur District, eastern Uttar Pradesh, India.

The 'Essential Grameen'

Whether we are providing training and technical assistance or establishing a new GBR, we adhere to what is now called the 'essential grameen'. It is the set of inter-related practices that are *necessary for a GBR to succeed in reaching and benefiting large numbers of the poor and poorest women and thereby making a significant impact on poverty, no matter what its context.* At the most general level there are only three essentials:

1. Exclusive focus on the poor, with priority for the poorest women;
2. Provision of financial services (small loans and savings facilities) in a way especially designed to facilitate their successful participation and timely repayment of the loans; and
3. Quickest possible attainment of institutional financial self-sufficiency that is consistent with the overriding goal of poverty-reduction

Exclusive Focus on the Poor and Priority for the Poorest Women

All international replications of GB that I know of, that started working mainly with poor men, like GB itself, have ultimately come to give priority to poor women.

Professor Yunus says this happened at GB because poor women proved to be better clients than poor men are, and any financial institution will give priority to its better clients.

Poor women proved to invest more of the loans in income generating activities rather than spending on consumption, they were more faithful at weekly repayment and they tended to spend more of the increased income on poverty reduction, especially for their children, than did poor men. Not only were they better clients, but also they were better poverty reducers. So naturally, once the Bank realised this, it gave priority to poor women in the expansion of its outreach to the poor.

But why are poor women better clients and poverty-reducers? Professor Yunus says it is because poverty is a women's issue. They feel poverty more deeply. They know first when there is not enough food; they know when their children have gone hungry, they hear them crying at night from hunger and watch helplessly, as their little bodies become bloated through malnutrition. They know better when their children need medical attention, and know why it cannot be sought. They watch the children of the non-poor putting on their clean, ironed school uniforms and going off to school everyday, while their own children, clad in rags, play in the dirt. They know poverty deeply and intimately, but they feel helpless to do any-thing about it. Usually they were born poor, and they don't dare dream of any other kind of life for themselves or their children.

Into this tragic but apparently unchangeable situation comes the Grameen Bank offering an opportunity for poor women to do something about their poverty and that of their children. A tiny ray of hope is ignited in their minds. Maybe their life can be better; maybe that of their children can be easier; and maybe they can bring it about themselves without depending on their husbands. So, despite being worried about getting into debt and not being able to repay, they force themselves to borrow but often only after seeing a poor neighbour succeed at it.

There is of course a flip side to this miracle story. Aminur Rahman (1999) who suggests, from his village-level observations in Bangladesh, that the Grameen Bank prefers women more for strategic reasons in relation to investment and recovery of loans than for the benefit of the women themselves has described it most fully, because they are more compliant and easier to discipline than the men. Moreover as the honour of their wives (and themselves) is at stake in repayment the husbands also pressure their wives to repay as required. Thus poor women are pressured from both sides, and some describe this as intolerable.

These observations are useful because these things do happen from time to time, in GB and probably most of its replications. Money lending, and fundamentally that is the core of the GB methodology for poverty-reduction, is not a pleasant business.

Always some borrowers will "test the limits" and try not to repay. Others will experience disasters, e.g., serious illness of husband, wife or children, death of the cow or buffalo purchased with the loan, flood or drought, etc., that make it impossible for them to repay as scheduled. GBR field staff, being pressed by their management to increase collection and to lower the portfolio at risk, and who may be paid incentives for doing so, may not bother to distinguish willful from unwillful defaulters, and may unfairly pressure the latter to repay. Ambitious branch and area managers may pressure their subordinate staff to maintain unrealistically high repayment rates, so as to maximise their chances for promotion.

In every large organisation there will be some bullies who will resort even to physical pressure and violence against subordinate staff and clients to get what they want. But well managed GBRs, that encourage upward flow of information, have strong field supervision and effective disciplinary procedures, can keep such excesses to a minimum.

For example, the Zonal Manager in Tangail, Bangladesh, where Aminur Rahman did his study, was recalled to Head Office for his excesses in 1995, and ultimately sacked for them. In the long run, however, it is the poor women them-selves, who will decide whether it is in their interest to continue in the program or not. It is essential for a GBR to have a cost-effective method of identifying and motivating poor women in their villages, in order to minimise the leakage to the non-poor, and for there to be adequate quality control over this process.

Ad 2. Provision of financial services to poor women in a way that facilitates their successful participation and timely repayment of their loans.

A Number of Important Policies and Practices Contribute to this Outcome

a) No collateral nor external guarantors required: It has been a major achievement of Professor Yunus to prove that banking does not have to be based on physical collateral nor external guarantors. GB's policy of not requiring such security for its loans has made it possible for hundreds of thou-sands of assetless poor women to avail of its financial services.

b) Delivery of the financial services to the villages of poor women: Rural women in Asia are tied to their villages by obligations to children and husbands, as well as, in some places, by culture and for safety. Moreover, poor women don't have the means to leave their villages. So if they are to be able to participate in a microfinance program its business must be carried out in their villages and in a simple way without complicated forms or procedures. Usually this is done through weekly centre meetings attended by a trained field staff of the program, but sometimes they are held every ten days or even bimonthly. While being an essential element of the GB methodology, this also has major implications for the cost of such programs, as will be seen below.

c) Formation of solidarity groups by the poor women themselves: This participatory process, which culminates in the federation of such groups into village based centres of around 30 poor women on average, gives dignity and collective strength to them, as well as the opportunity for leadership training.

   It also provides valuable information to the GBR on which poor households in the village are credit worthy, and enables it to impose collective responsibility on its borrowers for repayment.

d) Self-choice of loan activities : The women choose the loan activities themselves, based on their skills and experiences. Their group and centre must approve these, which then take collective responsibility for the loan repayment.

e) Small loans mainly for income generation: The loans are kept small, usually between US$50 and $75 initially, and only increase gradually for subsequent loans, as the ability of the poor to repay increases. The main responsibility for proper loan utilisation is placed with the group and centre, but field staff of the GBR also carry out thorough loan utilisation checks, one week after the funds are disbursed. Loans are disbursed in a staggered manner to group members in order to create credit discipline

f) Financial products designed especially for poor women: The financial products, both loans and savings, must be designed especially for the situation of poor women. Loan products must be adequate for the kinds of income generating activities in which they can engage. Savings products must take into account that deposits will be very small and withdrawals frequent.

g) Frequent repayment : Repayment is frequent, usually weekly but sometimes every 10 days or bimonthly, over a year at least for first loans in order to keep the installments small thereby making it easier for poor women to establish their creditworthiness.

h) Safety net: A small compulsory saving is deposited into a Group or Centre Fund at each centre meeting. This Fund is a source of emergency loans for groupcentre members, with the approval of their peers.

   It also enables them, over say five years, to buildup savings that can be used to purchase shares in the GBR which can, as has the Grameen Bank, become a community owned microfinance institution (MFI).

i) Eligibility for subsequent loans dependent upon repayment of the previous loans: Upon completing repayment of a loan, a client becomes eligible for a subsequent one, the maximum amount of which will be determined by her previous repayment record and her loan proposal as approved by group and centre. A poor household, not to mention the poorest, cannot come out of poverty with one or two loans. On average it may take five to ten small loans, over as many years, depending on the severity of the poverty in the area. Subsequent (possibly larger) loans, therefore, are vital for poverty reduction. They are vital also for attainment of institutional financial self-sufficiency by GBRs serving the poorest, as is seen below.

Quickest Possible Attainment of Institutional Financial Self-sufficiency that is Consistent with the Overriding Goal of Poverty Reduction

Reaching and benefiting large numbers of the poor and poorest women in their villages with financial services, even when done efficiently, requires large amounts of funds. For example, the Microcredit Summit target of providing financial services to 100 million of the poorest women is estimated to require about US$21 billion over the next 7 years.

The only possible *main sources* of funds of that magnitude are the savings and commercial financial institutions and markets. In order to be able to access such funds in the large amounts required MFIs (including GBRs) will have to become commercially sound institutions themselves; that is they will have to become profitable business concerns. It can be argued that this cannot be part of the 'essential grameen' because the Grameen Bank itself did not pursue such a strategy.

It is in its 16th year as a bank and still it is not financially self-sufficient as an institution. But it is making a significant impact on poverty in Bangladesh by providing financial services to more than 2 million poor women. Strictly speaking this is true.

While the Grameen Bank has been run in a business like way right from its establishment, it does not appear to have charged an appropriate interest rate to clients; that is, one that would cover all of its costs, particularly after they are adjusted for the effects of inflation and subsidies.

It is not, therefore, making real profits. Yet that has not prevented it from attracting the large amount of funds (probably around 300 million US dollars) it needed to reach more than two million poor households.

So why is it essential for smaller GBRs today to attain institutional financial self-sufficiency quickly, say within four or five years, in order to be able to make a significant impact on poverty in their countries?

Mainly because there are now many more of them and the targets and total fund requirements are much greater, but also partly because the world financial conditions have tightened and donors do not appear to be anywhere near as interested in the international replication of the Grameen Bank as they were in its expansion in Bangladesh.

From experience over the past eight years in CASHPOR, I can say that fund shortage has been the only insurmountable obstacle to much greater expansion of outreach of our member GBRs.

The following would appear to be essential in this process:

a) a three to five year business plan toward institutional financial self-sufficiency;

b) skilled financial, as well as field, managers;

c) increasing levels of institutional efficiency to emerging industry standards;

d) interest rates/fees to clients that are appropriate to cover all costs and to attract savings;

e) "near perfect" repayment,

f) a computerised management information system that produces financial statements at international standards;

g) an effective staff productivity incentive scheme.

The 'essential grameen' was identified first through analysis of the GB methodology, but it has been refined over the years mainly through trial and error. Something like the same process probably will have to take place with the PPHP.

The Art of Replication

The art of replication lies in the process of implementing the essentials in a different context. Functional equivalents have to be discovered for essentials that cannot simply be copied, because of cultural, economic or political barriers in the new context. This is the most creative and exciting part of the whole process.

It requires deep knowledge of the new context but also a meaningful degree of freedom from the psychological constraints that it imposes. To see this clearly let us look at the replication and adaptation of GB in Mirzapur District, eastern part of the Indian state Uttar Pradesh by CFTS. Because of limitations of space, we consider only differences between CFTS and GB on how the 'essential grameen' is implemented, that is the adaptations that CFTS had to make to implement the 'essential grameen' in its context.

Table: *Exclusivity for the poor and priority for the poorest women*

Grameen Bank	*CFTS India*
Uses agricultural land of less than 0.5 acre,.and total value of household assets less than that of one acre of average quality agricultural land in the area, to identify the poor on the ground. Does not distinguish between the very poor and moderately poor in targeting	Initially found it impossible to get accurate data on land ownership and operation because of suspicion among the poor. We used CASHPOR's House Index adapted to housing conditions in Mirzapur as a crude measure to identify the potential. Then as we wanted to distinguish between the very poor and the moderately poor, we added caste which is another easily observable characteristic in rural India, as in most villages housing is segregated by caste. So the very poor were scheduled caste or tribe households (SC&ST) with house scoring less than two points on our Index. The houses of the moderately poor score 2 or 3 points on our Index and tended to be in the other backward caste (OBC) section of the village.

In this case CASHPOR's House Index score and caste were used as functional equivalents of GB's land and asset criteria for identifying poor women in their villages. Later, when we were more trusted among the poor, we were able to get reasonably accurate information about land owned and operated, work of the husband and wife and possession of large farm animals. With that CFTS was able to sharpen its targeting and to be surer of the distinction between the very poor and the moderately poor.

***Table** : Designing the program to promote successful participation of poor women*

Grameen Bank	CFTS India
Most of the GB's clients do paddy husking with their first loan, especially poorest women. The first loan maximum is currently US$ 100 which is more than enough for paddy husking. It is a first loan activity particularly suited for the poorest because they have the skill and there is little risk, as there is an unlimited market.	In Mirzapur, there is no equivalent first loan activity to paddy husking. CFTS sets its first loan maximum at about US$ 57; but our clients tell us there is nothing they can do to bring in weekly income with that small amount. They want at least twice the amount to buy a milch buffalo. We are not prepared to risk that amount on unknown clients as we are worried about their repayment ability, particularly of the poorest. So they tend to be underrepresented among our clients.

In this case CFTS has not been able to develop a first loan product suitable for the poorest. Hence our keen interest in the PPHP. It looks like an ideal first loan activity for the poorest women, if it can be introduced successfully in Mirzapur District.

Grameen Bank	*CFTS India*
GB has basically only one loan product for its first-time borrowers, the General Loan with a term of 52 weeks. In the second loan cycle it offers both General and Seasonal Loans. The latter start out at half the amount of the General Loan, and have a term of 50 weeks.	CDTS started with only one loan product, its Initial General Loan (IGL) with a term of 50 weeks. Clients complained that not only was the loan too small but also they had to wait too long. So after discussion with them, we launched two new loan products for the first loan cycle: a savings-based loan which can be taken after six months, as a multiple of average weekly voluntary savings; and a balance based loan for those who did not borrow the full amount of IGL. Both additional loans require perfect repayment for the first six months and both have a term of 50 weeks.

By listening to its borrowers, CFTS was able to develop new loan products that enable its more capable clients to get ahead faster, but do not increase the risk in an unmanageable way, and increase the interest income of the Company.

Quickest Possible Attainment of Financial Self-sufficiency that is Consistent with the Overriding Objective of Poverty-reduction

As is mentioned above, it was not necessary for the Grameen Bank to adopt this strategy for expansion, but it has become necessary for its replications. Adopting it in India has proved to be quite controversial, although not so much among our clients. From the beginning CFTS has presented itself as a financial company with a social concern to reduce poverty in a sustainable manner. In the process we are charging our clients a flat rate of interest of 20% p.a. on their loans. This compares to the 20% p.a. on a declining balance that is charged by the Grameen Bank to its clients. So our effective interest rate of almost 40% p.a. is virtually double theirs.

Our spreadsheet analysis tells us that our current interest rate will enable us to break-even in our fourth year of operation and to cover our accumulated losses prior to breakeven by the end of the fifth year. We have been able to attract funding from the Grameen Trust only for the first three years. So clearly we cannot take longer to breakeven, which would happen if we lower our interest rate on loans to clients.

Nevertheless our interest rate has raised eyebrows among bankers and bureaucrats in India. They warn that the politicians will criticize us and force us to reduce our interest rate. So far this has not happened, but we are still small and not yet of interest to such people. It is our strategy to have made ourselves indispensable to our clients before the politicians pay much attention to us, and then they will have to accept us. Only time will tell if this was a good strategy.

Can the Poorest Pay?

This is the critical question of course. If the poorest can pay our required interest rate, and still have surplus income to begin to pull themselves out of poverty, then we have a win-win situation. If not, our strategy for attaining institutional financial self-sufficiency is inconsistent with our overriding general goal of poverty-reduction and our particular interest in making better the life of the poorest.

As we all know, in micro finance loans are small. Even relatively "high" interest rates on them still result in relatively small (in amount payable) installments, especially if these are paid frequently, say weekly, over a medium length loan term, say one to two years. In India a moderately yielding, say 4kg per day, milch buffalo can be purchased pregnant for around 4000 rupees (about US$100). If a loan of the whole amount is made available for that purpose to a very poor women at 20% interest (flat) for a term of 2 years with 100 equal weekly installments of principal and interest, each payment would amount to 52.8 rupees as per the following calculation (remembering that the 20% for the second year will be charged only on the outstanding balance at the year's beginning.

This is calculated to be 2400, as 32 rupees x 50 installments = 1600 rupees in principal would have been repaid in year 1. The interest on the balance of 2400 for year 2 will be 480 rupees added to the 800 rupees for year 1, this gives a total interest payable over two years of 1280 rupees and a total amount payable of 5280 rupees divided by 100 weekly installments this comes to an average of 52.8 per week (just over one US dollar).

The 4kg milk could be sold for 12 rupees a kg. This means that the weekly repayment money of 52.8 rupees could be earned in little over one day, leaving the income from the other 5 to 6 days to reduce the poverty of the household. The risk of the buffalo dying can be covered by livestock insurance at a premium of 4 rupees per 100 rupees of the animal's value per year or a total of about 160 rupees, which could be paid easily from the sales of the milk.

As the buffalo will produce milk for only about 9 out of 12 months, however, the client would have to save or engage in some other income-generating effort for the remaining three months. To fill the gap clients in India purchase a second buffalo as soon as they can. With two milch buffalo they can have a good steady income throughout the year, with which they can pull themselves and their families right out of poverty within a few years. A good example of this can be seen at the SHARE Branch in Dachepalhi, Gunter District, Andhra Pradesh, where more than half of the loans disbursed over the past 5 years have been for milch buffalo, and many of the original clients are now living in large concrete houses of their own design.

The buffalo example hints at a second important factor that makes it possible for the poor and the poorest to pay appropriate interest rates. The returns to capital in their microenterprises tend to average more than 100%. This was the finding of a recent impact evaluation study of CARD a Grameen replication in the Philippines by Mahabub Hossain (Hossain and Diaz, 1997). Returns to capital in his random sample of clients averaged 117%. As CARD's effective interest rate on loans to clients is about 39% p.a., this left 78% in the hands of clients to reduce their poverty. It can of course be argued that if CARD's interest rate were significantly lower, its clients could come out of poverty faster. But from where would they get their loans? If CARD does not charge an appropriate interest rate it would become bankrupt and would no longer be able to meet the financial needs of its clients. There is no certainty that another Microfinance Project (MFIP) would fill the gap. The only alternative for the poor may then be the traditional moneylender. A recent study of the returns to capital in micro enterprises in India and Kenya (Harper, 1998), found them to be even higher on average than did Hossain and Diaz at CARD.

The near perfect repayment rates, which are characteristic of MFIPs around the world, are empirical evidence that the poor can pay appropriate interest rates charged by efficient micro finance institutions.

Working in an area of India where repayment of IRDP loans is said to have been less than 10%, CFTS has been able to collect 96% of weekly repayments due, since it began operations 18 months ago. *There is no significant difference between the repayment rates of the poorest and the moderately poor clients.* CARD has maintained near perfect repayment for years, with about half of it clients coming from the poorest category. The sixteen CASHPOR-member MFIPs, who together had US$34 million in loans outstanding to over 200,000 poor and poorest households throughout Asia at the end of 1998, had a combined portfolio at risk of only 1.13%. The millions of weekly payments in full on time, that lie behind that figure are eloquent evidence of the ability of the poor and the poorest to pay appropriate interest rates for their financial services. So it is clear that the poor and the poorest can pay much higher effective interest rates on loans for income generation than has been thought.

Essential Conditions

The component activities of the 'essential grameen' have been listed above, and we have seen examples of how they are being implemented in Mirzapur District in India. To complete our analysis of the process of successful replication and adaptation, we must now identify any conditions, i.e., contextual factors, essential for its success.

Density of Poverty

The main costs of providing financial services to the rural poor are the administrative expenses (mainly salaries and allowances of the field staff), of delivering them to the village based centre meetings. These costs will increase if the poor households are geographically scattered, rather than living in villages in large-enough numbers to make possible the formation of a GB-style centre, with at least 6 groups (30 poor households). As the costs increase so also must the appropriate interest rate at which the micro credit should be provided, if it is to be a financially sustainable operation. There will be an interest level beyond which the poor women cannot earn a net profit from their microinterprises, after paying the principal and interest on their small loans. Once it is neared, they will no longer enter the program. This is a serious restriction, as the poorest households in a country often live in a scattered manner in its resource poor areas, e.g., in the stony mountains in southwest China or the hills and mountains of Nepal. Mention of mountains reminds that the nature of the terrain in which the poor live also is important. Nobody has yet successfully replicated and adapted the GB methodology in such difficult contexts. Fortunately most of the poor in China and Nepal do not live in such areas.

Freedom to Create Self-employment

The GBRs for the most part finance the creation of self-employment by poor women. If they are not permitted, say for cultural or political reasons, to engage in such activities, then this approach to poverty reduction could not succeed. It is hard to imagine a country in which women could be more restricted than in Bangladesh where the GB originated, however, so the cultural prohibitions would have to be very strict indeed to prevent poor women from taking advantage of the opportunities that micro finance offers. Probably this is not a serious barrier to the inter-national replication and adaptation of the GB Model.

Political restrictions on private enterprise were important barriers in some communist countries until recently, but this is no longer the case in those, like China and Viet Nam, where the largest numbers of poor households live. Should there be a return to centrally planned economies that forbid or severely curtail private enterprise in the rural areas, however, this would preclude successful replication and adaptation of the GB Model in such contexts. Other than the requirements of a geographical density of rural poor households that permits the formation of village based centres that contain at least 30 households, and the freedom of poor women to engage in and create self-employment, there do not appear to be any other essential conditions for the successful international replication of the GB Model. That is why we find it succeeding in many different national contexts all over the world.

Lessons for the Replication of other Development Programs for Poor Women

What lessons does our work in replicating the Grameen Bank have for replication of other development programs for poor women, such as the Poultry Production and Health Program? At the most general level, we should analyse the PPHP idea and the experience of implementing it in different national contexts to discover its 'essentials' that is, the minimum set of component activities and procedures that is necessary for its success in any context. At the same time, we should be looking for any essential conditions and contextual factors, for its successful implementation. Such analysis is beyond my expertise. What I can do, however, is to ask if any of the 'essential grameen', including the conditions essential for its success, appears to apply to the PPHP? Each of the three general components of the 'essential grameen' appears to be relevant to the international replication of PPHP.

Exclusive Targeting on the Poor and Priority for Poor Women?

At first sight this important lesson does not appear so relevant to the replication of PPHP because presumably it requires a minimum density of participants within a given locality in order to be feasible and efficient. If there were not enough poor women, then the non-poor would have to be given a chance. To ensure that the latter did not exclude the poorest women (likely in India because of caste prejudices), however, it would be a good idea to ensure that the poor and poorest women were given the first opportunity. For this they would have to be identified with a cost-effective targeting tool, like the CASHPOR Housing Index or Participatory Wealth Ranking.

It is the women and girls who will look after the poultry. Working with them will be more direct and it will improve their status. In most south Asian contexts poultry, especially of the scavenging kind, are usually seen to be the possession of the women. It is a fairly liquid form of savings for them. They can sell a hen when they need the cash, without prior permission of the husband. So by working primarily with women, the PPHP would be improving their economic position directly.

But why priority for the poorest women? Because they and their children are the neediest, and PPHP is a program that could make a literally vital difference for them. As mentioned above, outside of Bangladesh, it is difficult to find a suitable first loan activity for the poorest women. PPHP would appear to be just such an activity. It would require only a small investment, it would bring in additional income fairly quickly and regularly and it would entail minimum risk.

Design of PPHP to Facilitate the Successful Participation of the Poorest Women?

There would appear to be several parts of the PPHP for which this lesson is relevant. First, in the training of women. Most obvious are that the costs of the program should be kept within their reach and the training should be carried out in poor villages.

There should be no up-front fees which would prevent poor women from entering PPHP; and the training of poor women in PPHP should be carried out in their villages at times when they are relatively free - in months when there is little employment for agricultural labourers, for instance. Perhaps not so obvious, but very important, is that participation by poor women in the program must be voluntary. They must decide, after being sufficiently informed that it is a good opportunity for them. For this the motivation work and training materials will have to be suitable for illiterate women.

Attainment of Financial Self-sufficiency for PPHP as Soon as is Consistent with Poverty Reduction

If it is not designed to cover its operating costs from its operating income as soon as is consistent with the overriding goal of poverty reduction, PPHP will not be around long and will not be scaled-up to reach and benefit truly large numbers of poor women and their households. Poor women will have to pay appropriate costs for hens, chicks, essential veterinary services and food supplements. The timing and amounts of these payments should be planned carefully to ensure that most poor women would be able to make them.

Also the women will have to be convinced, by training and demonstration, that the payments are essential to their success with the program. Otherwise they won't make them.

Essential Conditions

A minimum density of poor households will also be an essential condition for PPHP, as will be freedom of poor women to engage in and create self-employment. In addition, an *efficient system for marketing the poultry products* must be in place. As there are already markets for eggs and poultry in most rural communities, the main problem may be to prevent oversupply that would drive prices below production costs. Probably it would be best to begin the program in areas, like Uttar Pradesh state in India, that are known to be poultry deficient.

Even this preliminary effort to find lessons in the international replication of the GB, that would be useful in spreading PPHP, has brought to our attention some potentially important matters. Identification of the 'essential PPHP', including conditions necessary for its success, is the obvious next step. After that, where necessary, functional equivalents will have to be found for implementing the 'essential PPHP' in different contexts.

Backyard Poultry System in the Tribal Areas

The State of Madhya Pradesh occupies an important place in tribal India. The tribal population in the state is about 10.2 million, which is nearly 23 percent of the total tribal population in the country (1991 census). The scheduled areas extend to nearly 13 districts, covering about 65,670 square kilometres. The basic problem of the tribals has been mainly poverty and exploitation.

The 'Gond' is a primitive tribe who has been primarily the residents of central India. Traditionally, the Gond practice cattle, pig, sheep and goat rearing and keep poultry as supplementary income. Due to poverty and exploitation, the tribe was almost going into oblivion. A Tribal Commission and a Tribal Welfare Board are constituted to give them protection and plan schemes for their improvement. More than fifty Gond subtribes are recognised. In the Shahpura block of Jabalpur district, about 105 small villages have a majority of Gond population, while the proportion is smaller in other villages. Their main occupation was 'hunting' and collection of forest produce. They specialise as 'Tendu Patta (leaves) collectors'. Due to restrictions imposed by Government, the hunting has reduced drastically.

This study is on the existing backyard poultry system with the objective to pave the way for development of backyard poultry into a sustainable income-generating activity for the tribal households.

Definition and Terminology

The Backyard Poultry (BYP) terminology as per this study includes the poultry rearing units in which the farm families have 1-10 scavenging adult birds feeding on broken grains, insects, kitchen wastes, green vegetables and leaves and anything edible available in the surrounding areas.

Importance of Study

Despite the large number of tribal households having BYP as a traditional practice few studies have been done on the subject in the Central or Eastern Madhya Pradesh and the backyard poultry has not been identified as a focus area in the tribal development programmers. However, according to the literature relating to development and the schemes being implemented BYP seems to have taken a backseat. The state veterinary department did implement a scheme wherein it distributed a few units of ten exotic birds each to a large number of the families, but this program has not been successful. There was yet another scheme in which 200 birds of the White Leghorn breed were being given per selected family, but this too did not succeed.

Why did these Government plans fail and are now given up when most of these tribal families do rear poultry? Thus, for any developmental intervention it was felt necessary to study the realities on the ground as existing in the area.

Objectives of the Study

Due to lack of literature, one of the objectives of my research work has been to study the existing system of BYP and the related problems and constraints experienced by the tribal families. The objectives of the study were as follows:

1. To study the present backyard poultry system in the Gond areas and their poultry rearing practices.
2. To study the constraints in the backyard poultry practices as listed by the tribal people and their possible solutions.
3. To suggest simple, low cost practices to enhance sustainable income generation through backyard poultry.

Limitations of the Study

There is not much literature available on poultry, which relates to areas where the Gond tribe predominates. BYP is unorganised in the Gond area and that makes it difficult to get reliable data. The sale and purchase of poultry products are ad hoc. It is therefore difficult to get authentic data on their marketing. Feed supplements to the birds are ad hoc making it difficult to study the nature and quality of intake of feed by the birds.

Methodology and Location

The study was conducted in Jabalpur district of Madhya Pradesh in central India. The Jabalpur district comprises 13 blocks of which two, i,e. Shahpura and Kundam have a large tribal population. Shahpura was selected for the study. Of the 105 tribal villages in Shahpura block, 12 villages were selected at random for the study. In these villages 54 per cent of the households had BYP. From the households having BYP 100 respondents were selected at random. The majority of the respondents, 46 percent, belonged to the Gond tribe, 14 per cent to the Bharia tribe and 40 per cent were from the scheduled caste and other backward classes as other households besides the tribals rear poultry. With the help of an interview schedule in-depth interviews were conducted with individual respondents. The data were systematically recorded, interpreted and analysed.

Findings

Total Income and Income from Poultry

The data indicated that 67 per cent of the respondents had an annual earning of up to Rs. 700 from poultry. 70 per cent earned up to Rs. 6000 from labour work (agriculture and construction) and 26 per cent earned up to Rs. 4000 from agriculture. The data indicated that as income from agriculture was low, there was a dependency on nonfarm activities for supplementary income. The climatic conditions are unpredictable and the landholdings becoming smaller with each generation and these factors put pressure on the rural families to look for other sources of income.

The analysis on the socioeconomic aspect of BYP indicated that at all levels of rearing, whether eggs, chicks or chicken, BYP has a strong potential as an income-generating activity in the tribal areas. The respondents indicated that majority of them earned reasonably from poultry. 45 percent estimated that they sell poultry produce only if there is an urgent necessity of cash. 33 per cent of the respondents expressed that earning from poultry was nil (as they do not consider the barter system as a source of earning), 50 per cent said that their earning was between 0.1 to 4.9 per cent of the total earning, 9 per cent expressed that it was between 5.0 to 8.9 per cent. The remaining 8 per cent earned between 9.0 to 14.9 per cent of the their total cash income from poultry. The families expressed that the income from poultry was meager and that they were afraid to rear poultry in larger numbers for the fear of an outbreak of an epidemic that would kill the entire poultry population.

Percentage of Households having BYP

It was found that the higher the percentage of tribal population in a village, the more households had poultry. The number of adult birds ranged from 2 to 10 with about 5-10 chicks. The villages Bothia, Somathy, Chargawan, Ahmedpur, Dongarjhansi, Kohali and Bijouri had on average 50-55 % households with BYP. Families with no backyard either had none or just 1-2 hens. Some villages like Piparia, Chhapra, Jamunia and Ghungri had about 35 % households with BYP. The rest were poor farm families, who either did not have backyard space or migrate for work. They sell their stock of birds prior to migration and acquire new ones upon their return.

Reasons for Chicken Rearing

76 of 100 respondents said that BYP could be a good source of income. 75 expressed that the birds are used for consumption during festivals and on special occasions. 9 respondents said that bird rearing was their hobby and interest. Only 2 respondents answered that the birds were for traditional rituals and sacrifices. This situation differs in the case of tribal households in Mandla district, also dominated by Gond tribes, where a large percentage of the rural households make use of poultry birds for traditional rituals and sacrifices.

Procurement and Sale

91 per cent of the respondents found that there was no problem in procuring or selling birds in the village itself. As far as the sales go, 83 per cent indicated that the avenues were within the village itself and 17 per cent sold to others. The selling price in the village per egg was Rupees 2-3, as expressed by 99 per cent of the respondents; Rupees 35-80 (according to 89 per cent of the respondents) per hen and Rupees 50-120 per cock as indicated by 82 per cent of the respondents. Nearly 16 per cent expressed a demand for cocks, 45 per cent said that there was a demand for hens, while the remaining 39 per cent expressed a demand eggs.

Preferred Poultry Breed, Egg Laying and Hatching

96 per cent said that their preference was for the indigenous poultry breeds. It was only 4 per cent, who preferred the exotic birds mainly due to their higher egg laying capacity. 8 per cent of the respondents said that their birds lay between 11-20 eggs/hen/year. 19 per cent that it was 21-30 eggs, 28 per cent had their hens laying 31-40, while 35 per cent said that their hens laid 41-50 eggs per year each. 9 per cent said their hens laid 51-60. The remaining 1 per cent had egg production of 61-70. 31 per cent hatched between 1-10 chicks per year, 59 per cent between 11-30 and 10 per cent had 31-40 chicks hatched per year. The respondents said that they generally do not sell the eggs and prefer to hatch them and rear the chicks, as this is more profitable.

Feeding and Annual Expenditure

82 per cent of the respondents feed their birds with broken grains as available in the season, but the feeding practices are very casual with no separate feeding for chicks and adults. Household left over food is given to the birds by 18 per cent of the respondents. It is difficult

to estimate the diet pattern of the birds because of the ad hoc supplementary feeding pattern. 76 per cent of the respondents said that they do not spend any money on supplementary feed (this meant that they give supplementary feed but did not purchase any). 24 per cent spent Rupees 50-500 annually on feed.

This ad hoc feeding system could be a major contributor to under nutrition and malnutrition, leading to unhealthy chicks and their early death.

Early Chick Mortality and Rearing

58 per cent said that the largest number of chicks die in the age group 1-10 days. 19 per cent said that the largest mortality was between 11-20 days. Once the chicks attain the age of 31-40 days, mortality is greatly reduced unless there is an outbreak of viral disease or an epidemic. The indigenous (Desi) birds are generally resistant to the parasitical diseases but not viral diseases. 62 per cent rear the chicks under a bamboo basket, 14 per cent put them in an almirah, 17 per cent have small mud chick houses and 7 per cent said that the chicks find a place for themselves inside the house along with the hen.

Awareness of Medication and Vaccination Programs

Though there is some awareness of the ailments and disease outbreaks in the birds, lack of veterinary services in the villages has led to no control on mortality. Thus, poultry is considered to be the most risky venture by the tribal households as any outbreak of disease or epidemic can wipe out the entire poultry population. The people were able to identify some common diseases, but have no remedy for their control or cure. 66 per cent of the respondents explained that their birds suffer and die of 'Kata' (i.e., Newcastle Disease), 4 per cent could identify 'Fafoondi' (fungus and toxicity) as the reason for mortality. 30 per cent expressed that they were not aware of and could not identify the diseases. There is a constant fear of disease outbreak and the entire flock getting wiped out. This fear prevents people considering poultry as a sustainable income-generating activity, although a large number of households do rear poultry birds.

General Problems

After the interviews and following the discussions with the tribal families the problems that have been observed are:

- Heavy mortality in chicks
  - Mortality in adult birds due to outbreak of disease

- Malnutrition in birds
  * Low egg production
  * Disease outbreaks and lack of veterinary assistance in the villages
  * Lack of awareness and knowledge about poultry practices.

Other problems were attack by predators (79 per cent), 64 per cent mentioned 'disease' as a major problem while 3 per cent mentioned non-availability of feed and medicine. 52 per cent felt that the adult birds are able to look after themselves and they do not need any special protection or security measures, while 17 per cent said that the birds find a place for themselves within the house in a basket or some secure corner. 31 per cent felt that the birds need personal care particularly to pre-vent attacks of wild cats and snakes.

Recommendations and Suggestions from the Study

There is a need for awareness building, training and systematic planning which will help develop BYP into a sustainable project for the upliftment of the rural population. Introduction of a dual-purpose bird with a black barred brown plumage (Krishna-J female x Synthetic male) with a higher genetic potential and resembling the 'Desi' (indigenous) birds in their physical characteristics, being developed by the scientists of the Agricultural University at Jabalpur, should be popularised with the farm families. This dual-purpose bird will have a triple advantage of the characteristics of the indigenous bird in terms of hardiness and colour, high egg laying capacity and high weight gain like those of the exotic birds.

This dual purpose bird will have an egg laying capacity of 120-130 nos. per year in the scavenging (free-ranging) system, and 220-240 nos. per year in the intensive system (characteristics from Krishna-J female) and attains a body weight of 1000 gms. in 6-8 weeks. The egg weight at 40 weeks age would be 50-51 gms having a tinted brown coloured eggshell, similar to that of an egg of the 'Desi' bird.

Since BYP is an arena of women, a planned effort should be made to develop BYP as a group activity by training a few women as vaccinators and some for manufacturing low cost feed formulations at the village level. If these resources were made available locally and with the synergy available in BYP, family poultry would contribute towards sustainable supplementary income to the tribal house-holds.

The findings of this study indicate that the maximum mortality occurs in the young chick stage. Simple management skills like covering the bamboo baskets with news paper or cowdung for insulation during brooding and protection in winters, creep feeding, low-cost balanced feed formulations, awareness about hygiene for feeding and watering of birds, burning of a lamp or charcoal burner inside the brooder in winter to give warmth to the chicks are some of the simple low-cost management practices suggested by this study to reduce early chick mortality.

The study emphasizes that with the synergy for BYP already existing in the rural areas and with minimum extra labour needed by the women despite their busy schedule, poultry has a possibility of success as a rural cottage industry. The precondition to that would be that a basic minimum medical facility is made available locally.

The results strongly supports the possibilities of establishing this presently neglected sector as a viable 'semi-organised' venture in the future by the national and international development agencies as an element of economic development in rural areas.

Poultry as a Tool in Poverty Eradication and Promotion of Gender Equality

Sixty professionals with experiences from developing countries in Asia, Africa and Latin America, employed by universities, research institutions, governments, NGOs, Consultancy companies and multi- and bilateral aid agencies met in Den-mark March 22–26, 1999 at the course centre Tune Landboskole to conduct a workshop on poultry as a tool in poverty eradication and promotion of gender equality. There were several good reasons that justified the workshop: Donor agencies and recipient countries alike have poverty alleviation and gender equally high on their lists of priorities. However, while a broad menu of productive interventions is difficult to identify that poor women profitably can undertake, the very positive impact on women's and children's lives recorded in impact studies of the semi-scavenging poultry model developed and applied in projects in Bangladesh served as a strong impetus, and it has inspired the creation in Denmark of a multi-disciplinary Network for Poultry Production and Health in Developing Countries. The Network has just had its first major grant approved by Danida and was interested in updating its knowledge on small scale rural poultry production as an input to detailed planning of its activities. On this background the objectives of the workshop were:

- to identify experience that can be replicated in other countries and projects

- to identify training and research needs
- to promote the development of policies and action plans towards these ends.

The workshop had representatives from several disciplines and received more than 30 papers and reports on topics dealing with the Danish human resource base in tropical agriculture; the situation in ongoing projects; scaling up, networking and replication; interactions between poultry breeds and their production environments, diseases and nutrition, capacity building (training and education) and links between users, institutions and technologies. In a final session the following topics were first discussed at length in group sessions and subsequently reported and commented upon in a plenary session:

Training and Information Dissemination

Training and Information Dissemination impinge on all the other topics under discussion, including the replication and adaptation of development models, research and extension and government policies. It is critical to the development of family poultry for poverty alleviation and gender equity. The subject is considered within the specific context (poultry, poverty and gender) but divided between training and information and subdivided into the needs of:

- farmers and family: men, woman and *children*
- extensionists and technicians (trainers)
- university or college teachers (trainers of trainers)
- researchers
- policy makers: local and national.

Farmers

Objectives

The objectives of training were considered to be:

- improved skills,
- increased income (poverty alleviation)
- increased food production and security
- empowerment
- exposure.

Methods

It is important to start from what the farmers know using interviews, discussions, etc. All the various methods were considered, including RRA, PRA, surveys and questionnaires. The various techniques of seasonal planning, demonstrations, farmer contact, tours, media (newsletters, radio, TV, theatre.) were included. The importance of monitoring and evaluation was also considered.

The *Integrated Farmer Field School* (IFFS) that was pioneered in Integrated Pest Management and now applied to other subject areas as well was considered the central model. It was described to the workshop in a presentation from Cambodia and it contains all the elements of the methodology and subject matter (with 70% practical work, village group formation and group work).

But it was noted that the BRAC/DLS model in Bangladesh that almost exclusively serves landless women differed in the duration of training and the degree of formality.

Training was 4-5 days followed by return to the household, then a refresher course, compared to 4 hours per week for 20 weeks the first season in the IFFS.

The differences between subject matters, countries, gender (women's responsibilities in cooking require attention to hours of the day and location of the training) and cultures (women's mobility, previous exposure to education and degrees of communication between household members, i.e. does the trained person communicate his or her skills) in preparation of the training schedule were noted.

But the training still reaches *too few people* and is *too costly in time and resources*. One answer is to *train more trainers*. The training of farmers by farmers and selling their training skills was one long-term goal. Another solution is levying of technical service charges as in the programmes in Bangladesh that apply microcredit.

Trainers

One major problem, apart from budget constraints, is that extensionists and technicians are too tied up with their work and their time is limited, the information that reaches them is scarce since there is no flow of information from the regional, national or international level, and much of the information is out of date or not relevant as the subject of poultry as a tool in poverty alleviation has been and is neglected by research. This situation needs to be improved.

This could be by the *dissemination of relevant information* material that could be used in farmer manuals, which includes:

- vaccination manuals
- creep feeder designs
- housing

- brooding
- rearing and so on ... The material could be produced on paper, video, audio, etc.

Problems include the language and relevance to the users, so a manual should be for two levels: one at the level of the trainers and one for the farmer. The importance of pictures was stressed, i.e. *"talking pictures"*.

A comprehensive manual for trainers was referred to as an *idea catalogue, which* can be modified, subject to the place and people.

It was discussed who should be responsible for constructing the *idea catalogue*. It was concluded that everyone could contribute and participate in a global *website*. This would include all the elements above, especially the *talking pictures* and particular attention was needed to *relevance* to the needs of trainers and farmers.

The information technology aspects were not discussed as it was concluded that the facilities existed on the *world wide web* and the possibilities of computer communications, CD-ROMs, etc. could be used by all and were becoming available in all countries as witnessed by the near 100% access by the workshop participants. However, there is a need for donors and governments to enable the outreach and quality of access.

The idea catalogue would allow exchange of ideas and experiences between ecoregions and countries. Information is disseminated locally at traditional centres: coffee houses, places of worship, bars and other common meeting points. In principle the *cybercafé* falls into the same category.

Refresher courses are also needed. One time training is not enough. There must be *follow up* and the *long term goal* must be that training is done by farmers or their organisations, that they can *sell* their knowledge and therefore be independent of the initial help that needs to be given to them.

Funding for training must come from the farmers, i.e. they pay for training, so that there will be motivation. In Mali for example, farmers developed their own consulting group. NGOs could be created by farmers, probably not called co-operatives as these are not liked by farmers in some countries, but other examples were presented to the workshop like NGOs in Bangladesh moving towards financial independence and the Grameen Bank replication in Mirzapur, India that aim at financial independence within five years and yet plan to include a poultry program.

Who will pay for training the trainers? Initially one has to start with donor funds and then lead them into self-dependent processes. Donors need to start, but it must eventually be sustainable.

Training of Trainers

This has been the weak link in the process. They need training in both methodology and subject matter.

Most extensionists and technicians are trained in universities and colleges, but these institutions do not focus on the needs of trainers and, specifically for rural poultry, this has not been included in the curriculum. It was discussed whether there was a need for new institutions or whether existing institutions could adapt to the need. The problems of changing existing institutions were recognised but this was not ruled out.

However, it was not suggested that specialist poultry institutes should be set up as the problem applies to rural livestock, sustainable natural resource use systems and integrated farming as well. Besides experience does already show that while an initial investment in poultry for very poor people represents a first step out of poverty, a subsequent investment may be in other activities such as other types of live-stock, farming activities, land or nonfarm activities.

It was discussed how this training should be done and it was found that the proceedings of the workshop would contain useful examples from the models presented. These include for *farmers* the IFFS in Cambodia and BRAC/DLS in Bangladesh, the client focused training of *extensionists* by the Nordic Agricultural Academy and the integrated natural resource utilisation model for *graduate and postgraduate students* by the University of Tropical Agriculture Foundation in Vietnam and Cambodia. A common feature in these models is their focus on practical application of the skills obtained in training. In fact practical on-farm work with farmers is a common denominator.

Researchers

Research in the area of scavenging and family poultry is almost nonexistent. There is therefore an urgent need to involve researchers at all levels of scavenging and family poultry production systems.

Students should identify their own projects and there should be more participation, more on-farm research and feedback to the farmers' needs. Several references during the workshop were made to one experiment in which about 300 poor women acted as experimental hosts in a poultry breed evaluation, but it was also shown in another paper that there is a strong need for researchers to broaden their methodological tool kits for participatory research with farmers to be meaningful and to link logically between problem identification, research design and conduct of the research.

It was suggested that there should be a kind of community service or 'military service' for all researchers to ensure that they remain in touch with the farmers' problems.

Policy Makers

Data must be collected and provided to the policy makers to convince them of the importance of rural poultry as a tool in poverty alleviation and gender equality and the associated training needs. At local and national levels, poultry must be included in budgets and legislation and obstructions of subsidies and taxes need to be removed. A liberal market condition should be developed and infrastructure should be generally improved to facilitate access to markets.

Conditions for Replication and Adaptation

There was disagreement with regard to the possibilities for replication and adaptation and the variety of opinions ranging from:

> *"the idea of replication is absurd and should not be discussed"*
>
> *"anyway it cannot be done, it is not possible, because conditions are never the same in a different context"*
>
> *to the view*
>
> *"replication has already been done in the case of microcredit".*

It was pointed out that microcredit was an exception, but for scavenger poultry, replication of the Bangladesh (BRAC/DLS) model in other countries does not seem to be possible although it is noted that there are important lessons in the emphasis the Bangladesh NGO BRAC's top managers put on research and training as a precondition for expansion of its programmes.

In place of replication and adaptation it was discussed how to improve scavenger poultry production in different contexts.

Consensus was reached to list some practices based on professional experiences, which emphasize the following points:

- Improvements need to be location specific. For example in Burkina Faso, where poultry raising is common, inputs in terms of labour, feeding and healthcare are low. Therefore, management attitudes have to be changed. Supplementary feeding would have to depend mainly on locally available resources

- The objectives are to improve productivity and reduce mortality of family poultry using low management inputs
- Periurban areas may be particularly suitable for improving scavenger poultry according to the Bangladesh model, due to better infrastructure and a higher population density. However, they may require their own approach
- Improved breeds are of secondary priority to reductions in mortality and improved nutrition of traditional birds
- Credit may be a necessary component, but it is likely to be supplied in different ways for different countries
- Focus should be on the poorest villages and it should be kept in mind that the factors, which determine success, could be both human and animal. Scavenger poultry can be seen as opportunistic harvesting of a wildlife population (low input)
- "Money talks" or quick cash impresses villagers: Therefore training should be practical, given in short segments and should quickly result in increased income
- Organisational improvements should be long-term commitments
- Before any intervention, baseline (technical and socioeconomic) assessment, preferably participatory should be conducted
- Extension should pay frequent visits in the village
- Participatory, on-farm research with voluntary farmers is recommended
- Monitoring to gather qualitative and quantitative data
- One ethical component of on-farm research that it is crucial to know about is how to compensate farmers of a negative control group
- Training needs assessment.

Research and Extension

The following topics were dealt with:

1. Research areas
2. Research capacity
3. Research approach
4. Dissemination and publication

5. Research and extension links
6. Indigenous knowledge.

Research Areas

Smallholder poultry production aimed to benefit poor women and their families has never been a priority of the international (CGIAR) agricultural research system or the National Agricultural Research Systems.

In general the amount of research conducted has been very little on any of the relevant subjects be it diseases, socioeconomic (gender, culture, income or production systems), management including hatching practices and housing, marketing and processing, monitoring and evaluation and indigenous knowledge. In the case of nutrition and genetics some more may be known, but far from enough. Areas of immediate concern to research were identified to be:

Diseases

There is little information about diseases (including nutritional diseases). Baseline studies involving long term studies have to be done. This should include postmortems and isolation of diseases agents. The relationship between marketing practices and diseases need to be studied.

Nutrition

Supplementary feeding based on locally available resources that accounts for the natural environment and seasons. Such work should be combined with breeding to study any interactions between nutrition and breeding.

Genetics

First collection of local breeds is required and then follows identification of eco-types. The third step is studies on some of the identified eco-types, but to yield meaningful results such studies must be conducted in the system of production (scavenging or semi-scavenging), where the results will have to be applied.

Socioeconomic

There is a need for research that uses data disaggregated by age, sex and socioeconomic group in order to know within a household who does what. Who controls which resources, the division of labour, cultural aspects, etc. The contribution of poultry within the context of the household's farming system is little understood.

Management

The need is for descriptions and comparisons of systems of management to identify areas for research and extension on topics like general care, hatching and housing.

Marketing and Processing

There is a need for surveys that describe market channels and identify existing and potential bottlenecks within them - and their contribution to spread of diseases.

The workshop also identified a need to understand better, which products (eggs, meat, chicken raising, and barter) are most important under a given set of circumstances.

Monitoring and Evaluation

The main need is for development of qualitative methods of monitoring and evaluation and there is a need to find indicators that can be used in such evaluations.

Judged from the report on replication of the Grameen Bank presented to the workshop, there may be important features in existing monitoring systems applied on other subject matters to incorporate in the poultry work.

Research Capacity

In view of the neglect so far, there is an obvious need for building up national and international research capacity on smallholder poultry production aimed to benefit poor women.

Research Approach

The dominant approach should be holistic, multidisciplinary and participatory. Obviously, the research approach (es) chosen in a particular situation will depend on the topic(s) to be researched, but the emphasis should be on providing the scientists with a sufficiently broad menu of methodologies that will allow import-ant questions of relevance to poor women to be researched, and which will avoid that limited knowledge of methodologies leads to rejection of important research questions.

These considerations are particularly important in situations of on-farm and participatory research. However, research should be done at all levels according to the needs (on farm, on station and in laboratory) at local, national and international (ILRI and other organisations) levels.

Dissemination and Publication (Academic Level)

Judged by the previous record, scarce funding and limited expertise, smallholder poultry is not yet regarded as an area of importance in terms of political aspects or scientific prestige.

Publication

One reason that may contribute to the low importance of rural poultry in terms of political and scientific prestige is a general lack of literature showing the importance of rural poultry.

Researchers are therefore urged to carry out research in this area and to publish the work in international refereed journals. Some problems were identified in relation to this:

1. Quality of the current work being carried out
2. Lack of suitable journals (do we need to create our own journal?)
3. Existing journals may not accept papers on scavenging poultry, which may relate to the perceptions of science of the referees of international journals as many problems are of a multidisciplinary nature.

Dissemination

Several proven tools exist for dissemination of information.

The ongoing first electronic conference on family poultry sponsored by the International Network for Family Poultry Development (INFPD) and FAO clearly illustrates that this is one tool for exchange of information and INFPD now publishes its newsletter electronically as well as in hard copy by airmail and it is one obvious publication to use for dissemination.

Workshops and international conferences are proven tools and important for initial face-to-face contacts.

Popular articles can be written for newspapers and magazines and programs prepared for radio and television.

Research and Extension Links from the Extension Perspective

The limited outreach of livestock extension across developing countries is well known. The experience from Bangladesh shows that in a given situation a strategy combining microcredit, targeting of poor women and NGOs can provide a very effective outreach mechanism beyond the limited number of farmers having large ruminants that government livestock extension services traditionally confine them-selves to. As the dominating policy mood continues to be for slimmer government and privatisation, there is presently much experimentation with alternative extension mechanisms taking place and research and experimentation in this area need to continue.

There is still a strong need to make research relevant through farm orientation and use of applied and participatory methods.

Regular workshops, close relationship between researchers and extension workers, feedback from extension workers to researchers and vice-versa must continue to be encouraged.

Indigenous Knowledge

There is a need to study indigenous knowledge in relation to scavenging poultry. One obvious area could be the knowledge of indigenous plants in the treatment of diseases.

Government Policy on Poverty and Gender

There must be commitment to poverty eradication and gender equality and the governments shall provide the legal background for NGOs and other organisations to contribute to poverty eradication and gender equality in line with the international declarations such as for poverty eradication:

The Dhaka Declaration of the 7th SAARC summit of April 1993, the Social Summit held in Copenhagen, Denmark, in March 1994 and the Microcredit Summit, Washington D.C., February 1997.

According to these declarations the national governments should have specific policy instruments, strategies and legislation to reduce poverty among children, disabled and the hard core poor, which should be reflected in their national development plans, budgets and institutional structures. There should be adequate human resources dedicated to poverty eradication (e.g. 1 extension officer per 500 poor) and openness to development activities by NGOs.

These concerns should also be reflected in the priorities of the development agencies.

There should be mass mobilisation of the disadvantaged by NGOs coupled with cooperation (lobbying) of government functionaries and closer GO-NGO collaboration. Transparency and accountability should prevail at all levels.

For other disadvantaged rural groups such as the disabled: the blind, lame, deaf, crippled and emotionally disabled, etc. who make up part of the hard core poor, the national governments should legislate in their support. Such legislation may comprise targeting of income generating activities at the disabled and disadvantaged, using subsidies if necessary and government should ensure that established social welfare systems are reactivated and strengthened.

For women:

Pertinent declarations are the International Declarations for Women and the Beijing Declaration of 1995. According to these the national governments should main-stream gender concerns and make special provision in favour of women.

This is all the more justified as it has been conclusively documented that poor women are more committed to use development projects for the improvement of their families' welfare. They use increased incomes for family food security and the family's economic advancement.

Governments should therefore legislate for gender equality and ensure enforcement of the laws and make the necessary constitutional provisions to favour women (e.g. article 28 of the Bangladeshi Constitution states: "Women shall have equal rights with men in all spheres of state and public life". The article states further "Nothing in the article shall prevent the State from making special provision in favour of women and children or for the advancement of any backward section of citizens").

Education should be free and compulsory up to the first 10 years of school and provision should be made for scholarships to girls. Child marriages should be prohibited (minimum 18 for girls and 21 years for boys) and there should be legislative control of polygamy and polyandry. Fairness in inheritance laws should be ensured.

There should be legislation in support of other disadvantaged rural groups such as the blind, lame, deaf and crippled, who make up part of the hard core poor.

Income generating activities should be targeted at the disabled and disadvantaged; using subsidies if necessary and established social welfare systems should be reactivated and strengthened.

Family Poultry Development should be seen as a tool in the above socioeconomic context and governments, bi- and multilateral development agencies and institutions and NGOs should place more emphasis on family poultry in order to reach their target groups of women and the poor. They should establish a policy of Family Poultry Development as a means of income generation, poverty eradication, and employment and protein food production.

Family Poultry is accessible to the poorest of the poor as proven once again during the present workshop. It enables the illiterate to develop their skills and knowledge, which could be applied to other activities as well. It is a means to human resource development.

National governments should ensure that inputs (day old chicks, vaccines, diagnostics and feed analytical services and research support) are provided not necessarily by undertaking the activities, but as a minimum by providing conducive policy frameworks.

- Which is managed by Dr. E. Fallou Guèye, who participated in the workshop.
- This was the topic of the 1997 Tune workshop and the proceedings are available on the Internet at this address:

List of Chicken Breeds

There are hundreds of chicken breeds in existence. Domesticated for thousands of years, distinguishable breeds of chicken have been present since the combined factors of geographical isolation and selection for desired characteristics created regional types with distinct physical and behavioural traits passed on to their offspring.

The physical traits used to distinguish chicken breeds are size, plumage colour, comb type, skin colour, number of toes, amount of feathering, nipple (areola) colour, egg colour, and place of origin. They are also roughly divided by primary use, whether for eggs, meat, or ornamental purposes, and with some considered to be dual-purpose.

In the 21st century, chickens are frequently bred according to predetermined breed standards set down by governing organisations. The most commonly used of such standards is the *Standard of Perfection* published by the American Poultry Association (APA), the oldest livestock organisation in the New World. Others include European standards (especially British ones), and that of the American Bantam Association, which deals exclusively with bantam fowl. Only some of the known breeds are included in these publications, and only those breeds are eligible to be shown competitively. There are additionally a few hybrid strains which are common in the poultry world, especially in large poultry farms. These types are first generation crosses of true breeds. Hybrids do not reliably pass on their features to their offspring, but are highly valued for their producing abilities.

By Place of Origin

Australia

- Australorp
- Australian Langshan

- Australian Pit Game
- Australian Game.

Austria

- Altsteirer
- Sulmtaler.

Belgium

- Barbu de Watermael
- Belgian Bearded d'Anvers (or Antwerp Belgian)
- Bearded d'Uccle
- Belgian d'Everberg
- Campine
- Brabanter
- Braekel (Brakel).

Brazil

- Balazço Cealio
- Brazilian Grey
- Kalabèo.

Bulgaria

- Black Shumen chicken
- Starozagorska red chicken.

Canada

- Chantecler
- Red Shaver.

Chile

- Araucana.

China

- Cochin
- Croad Langshan
- Nankin
- Pekin
- Silkie.

Croatia

- Croatian Dwarf chicken (Hrvatska patuljasta kokoš)
- Dalmatian chicken (Dalmatinska kokoš)

- Hrvatica (Hrvatica)
- Križevac Crested chicken (Križevaèka kukmasta kokoš)
- Sava Crested chicken (Posavska kukumasta kokoš)
- Slavonian Dwarf Naked Neck chicken (Slavonska patuljasta golovrata kokoš).

Cuba

- Cubalaya.

Czechia

- Czech gold brindled hen (Èeská slepice zlatá kropenatá, Èeská zlatá kropenka, Èeška)
- Šumavanka.

Egypt

- Egyptian Fayoumi.

Finland

- Finnish Chickens.

France

- Bresse
- Bourbonnaise
- Crèvecœur
- Faverolles Houdan
- La Flèche
- Marans.

Germany

- Annaberger chicken
- Augsburger chicken
- Bergischer Long Crower
- Bergischer Schlotterkamm
- Bielefelder
- Deutscher Sperber
- Deutsches Reichshuhn
- Dresdner chicken
- East Frisian Gull
- German Faverolles

- German Langshan
- Hamburg
- Kraienköppe (Twentse)
- Lakenvelder
- Niederrheiner chicken
- Pfalz chicken (Pfälzer Kampfhuhn)
- Phoenix
- Ramelsloher
- Rheinlander
- Saxonian Chickens
- Sundheimer chicken
- Thuringian Bearded chicken
- Vogtländer chicken
- Vorwerk
- Westphalian chicken.

Greece

- Alonissos island Chicken
- Boufunes
- Curly Chicken Mutation
- Follidotes Chicken
- Greek Cuckoo spotted chicken
- Hooded Greek Chicken
- Chios Fighting Chicke
- Kalamata Chicken
- Komotini Long Crow Chicken
- Lesvos Dwarf Naked Necked Chicken
- Lesvos Fillianes Chicken
- Milos Island Chicken
- Pomak Fighting Chicken.

India

- Giri Raja (The Mountain King)
- Kalinga Brown
- Mumbai Desi
- Grama Lakshmi

- Naati Kori(Kudla)
- Kadaknath
- Dwarf
- Control Broiler
- Aseel
- Dalhem Red
- Vanaraja (The Forest King)
- Gramapriya
- Punjab Broiler
- Brahma
- Nicobari.

Iceland

- Icelandic chicken.

Indonesia

- Ayam Cemani
- Bekisar (interspecific hybrid)
- Sumatra (chicken)
- Kedu (nationally standardised)
- Nunukan/Tawao
- Pelung (long-crower, locally standardised)
- Ketawa ("laughing" or staccato-crower, local bred from Rappang, South Sulawesi)
- Bali (naked-necked chicken).

Iran

- Manx Rumpy (or Rumpless Game)
- Orloff.

Italy

- Ancona
- Leghorn
- Sicilian Buttercup.

Japan

- Japanese Bantam (or Chabo)
- Shamo (or Ko-Shamo)
- Tomaru

- Totenko
- Onagadori.

Korea

- Gangwon Jaeraedak
- Jangmigye
- Yeongsan ogye
- Han Do.

Kosovo

- Kosova Long Crowing Rooster.

Malaysia

- Malay
- Serama.

Marianas

- Saipan Jungle Fowl.

Netherlands

- Barnevelder
- Booted Bantam
- Dutch Bantam
- Hamburg North Holland Blue
- Polish
- Welsummer.

New Zealand

- Bawu Hawu
- New Zealand Junglefowl
- Waki Waki Hawa.

Norway

- Norwegian Jærhøne.

Pakistan

- Asil (or Aseel)
- Afghan game fowl
- Buff chicken.

Philippines

- Philippine Native Chicken.

Poland

- Green-legged Chicken
- Poland.

Portugal

- Pedrês Portuguesa
- Amarela
- Preta Lusitânica.

Romania

- Transylvanian Naked Neck.

Russia

- Orloff
- Yurlov Crower.

Serbia

- Banat Naked Neck (Banatski gološijan)
- Sombor chicken (Somborska kaporka)
- Svrljig chicken (Svrljiška kokoš).

Slovakia

- Oravka.

South Africa

- Ovambo
- Potchefstroom Koekoek
- Venda.

Spain

- Andalusian
- Asturian Painted Hen
- Castilian
- Catalana or Prat Leonada
- Empordanesa
- Euskal oiloa
- Extremaduran
- Flor d'Ametller
- Ibiza
- Indio de León

- Majorca
- Minorca
- Murciana

- Pardo de León
- Pedresa
- Penedesenca
- Pintarrazada
- Pita Pinta Asturiana
- Serrana de Teruel
- Sobrarbe
- Spanish game
- Sureña
- Utrerana
- White-Faced Black Spanish.

Switzerland

- Appenzeller (Barthühner and Spitzhauben)
- Schweizer chicken (Schweizerhuhn).

Turkey

- Sultan
- Hint Horoz
- Gerze.

Ukraine

- Poltava.

United Kingdom

- Derbyshire Redcap
- Dorking
- Indian Game (or Cornish)
- Ixworth
- Madly Blue
- Marsh Daisy
- Modern Game
- Muffed Old English Game
- Norfolk Grey

- Old English Game
- Old English Pheasant Fowl
- Orpington
- Rosecomb
- Scots Dumpy
- Scots Grey
- Sebright
- Sussex.

United States

- Ameraucana
- American Game
- Buckeye
- Blue hen of delaware
- California Gray
- California White
- Delaware
- Dominique
- Holland Iowa Blue
- Java
- Jersey Giant
- Lamona
- New Hampshire
- Plymouth Rock (or Barred Rock, Rock)
- Rhode Island Red
- Rhode Island White
- Winnebago
- Wyandotte.

Vietnam

- Ac
- Ga Noi
- Ga Tre.

By Primary Use

All chickens lay eggs, have edible meat, and possess a unique appearance. However, distinct breeds are the result of selective breeding to emphasize certain traits. Any breed may technically be used for general agricultural purposes, and all breeds are shown to some degree. But each chicken breed is known for a primary use.

Eggs

Many breeds were selected and are used primarily for producing eggs, these are mostly lightweight birds whose hens do not go broody often.

- Ameraucana
- Ancona
- Andalusian
- Araucana
- Asturian Painted Hen
- Barnevelder
- Campine
- Catalana
- Easter Egger
- Egyptian Fayoumi
- Norwegian Jærhøne
- Kraienköppe (Twentse)
- Lakenvelder
- Leghorn
- Marans
- Minorca
- Orloff
- Penedesenca
- Sicilian Buttercup
- White-Faced Black Spanish
- Welsummer.

Meat

Some breeds are preferred for meat alone, though the commercial broiler market is currently monopolised by the Cornish-Rock (a hybrid of the Cornish and Plymouth Rock). Many smaller farms and homesteads use dual-purpose breeds for meat production.

- Bresse
- Indian Game (or Cornish Game)

- Ixworth
- Jersey Giant.

Dual-purpose

The generalist breeds used in barnyards the world over are adaptable utility birds good at producing both meat and eggs. Though some may be slightly better for one of these purposes, they are usually called *dual-purpose* breeds.

- Australorp
- Brahma
- Braekel (Brakel)
- Buckeye
- California Gray
- Chantecler
- Cubalaya
- Derbyshire Redcap
- Dominique
- Dorking
- Faverolles
- Holland
- Iowa Blue
- Java
- Jersey Giant
- Marsh Daisy
- Naked Neck
- New Hampshire
- Norfolk Grey
- Orpington
- Plymouth Rock
- Poltava
- Red Shaver
- Rhode Island Red
- Rhode Island White
- Scots Dumpy
- Scots Grey
- Sussex

- Winnebago
- Wyandotte.

Exhibition

Since the 19th century, poultry fancy, the breeding and competitive exhibition of poultry as a hobby, has grown to be a huge influence on chicken breeds. Many breeds have always been kept for ornamental purposes, and others have been shifted from their original use to become first and foremost exhibition fowl, even if they may retain some inherent utility. Since the sport of cockfighting has been outlawed in the developed world, most breeds first developed for this purpose, called game fowl, are now seen principally in the show ring rather than the cock pit.

Bantams

Most large chicken breeds have a bantam counterpart, sometimes referred to as a *miniature*. Miniatures are usually one-fifth to one-quarter the size of the standard breed, but they are expected to exhibit all of the standard breed's characteristics. A *true bantam* has no large counterpart, and is naturally small. The true bantams include:

- Belgian Bearded d'Anvers
- Belgian Bearded d'Uccle
- Belgian d'Everberg
- Booted Bantam
- Dutch Bantam
- Japanese Bantam
- Nankin
- Pekin
- Rosecomb
- Sebright
- Serama
- Silkie.

Cross-breeds

Many common strains of cross-bred chickens exist, but none breed true or are recognised by poultry breed standards. Thus, though they are extremely common in flocks focusing on high productivity, cross-breeds do not technically meet the definition of a breed. Most cross-breed strains are sex linked, allowing for easy chick sexing.

- Black Sex Link (also called Black Stars)
- Red Sex Link (also called Red Stars)
- ISA Brown
- Lohmann Brown
- Daisy Belle
- Cream Legbar
- Cornish-Rock
- Easter Egger
- Broiler.

List of Duck Breeds

The domestic duck, like other poultry species, has many breeds. Most are derived from the wild Mallard, while a minority are descendants of the Muscovy Duck. Duck breeds are officially recognised for showing by governing bodies of enthusiasts such as the American Poultry Association, which usually administer breed standards such as the Standard of Perfection.

List of Duck Breeds

- Abacot Ranger (also known as Streicher)
- Allier Duck (Blanc d'Allier)
- Ancona duck
- Aylesbury Duck
- Bali Duck
- Black East Indian
- Blue Swedish duck
- Buff Orpington Duck
- Call Duck
- Challans
- Cayuga Duck
- Chara Chemballi Duck
- Crested Duck
- Danish Duck
- Duclair
- Dutch Hookbill (*kromsnaveleend*)
- East Indie Duck
- Forest Duck (Eend van Vorst)

- Gimbsheimer
- Golden Cascade
- Gressingham Duck (Wild Mallard crossed with Pekin)
- Huttengem Duck
- Indian Runner Duck
- Khaki Campbell
- Loon
- Magpie Duck
- Majorcan Duck
- Muscovy duck
- Orpington duck
- Pekin Duck (also known as Long Island duck)
- Philippine Duck
- Rouen Duck
- Saxony Duck
- Semois
- Silver Appleyard Duck
- Silver Bantam Duck
- Termonde Duck
- Venetian Duck (Germanata Veneta)
- Welsh Harlequin Duck
- Wood Duck
- Swedish Duck
- Swedish Yellow Duck.

Other:

- Australian Spotted (origin disputed)
- Barbary Duck (Muscovy-derived)
- Cayuga Duck (American Black Duck-derived)
- Mulard duck (Barbary drake crossed with mallard-derived hen; sterile).

List of Goose Breeds

This list contains breeds of domestic geese as well as species with semi-domestic populations. Geese are bred mainly for their meat, which is particularly popular in Germanic languages countries around

Christmas. Of lesser commercial importance is goose breeding for eggs, schmaltz, or for the fattened liver (*foie gras*). A few specialised breeds have been created for the main purpose of weed control (e.g. the Cotton Patch Goose), or as guard animals and (in former times) for goose fights (e.g., the Steinbach Fighting Goose and Tula Fighting Goose).

Goose breeds are usually grouped into 3 weight classes: Heavy, Medium and Light. Most domestic geese are descended from the Greylag Goose (*Anser anser*). The Chinese and African Geese are the domestic breeds of the Swan Goose (*A. cygnoides*); they can be recognised by their prominent bill knob.

Some breeds, like the Obroshin Goose and Steinbach Fighting Goose, originated in hybrids between these species (the hybrid males are usually fertile). In addition, teo goose species are kept as domestic animals in some locations, but are not completely domesticated yet and no distinct breeds have been developed.

Breeds

- Adler Goose
- African Goose
- Alsatian Goose
- American Buff Goose
- Aonghus Goose
- Arzamas Goose
- Bavarian Landrace
- Benkovska White Goose
- Brecon Buff Goose
- Celle Goose (*Celler Gans*)
- Chinese Goose
- Cotton Patch Goose
- Czech Goose (*Èeská husa*)
- Czech Tufted Goose (*Èeská chocholatá husa*)
- Danish Landrace
- Drava Goose(Dravska guska)
- Diepholz Goose (*Diepholzer Gans*)
- Embden Goose (*Emdener Gans*)
- Emporda Goose (*Emporda-Gans*)
- English Saddleback Goose

- Euskal Antzara Goose
- Faroese Goose
- Franconian Goose (*Fränkische Landgans*)
- German Laying Goose (*Deutsche Legegans*)
- Gorky Goose
- Greyback Goose
- Hungarian Goose
- Italian Goose
- Kartuzy Landrace (*Kartuska*)
- Kholmogorsk Goose
- Kielce Landrace (*Kielecka*)
- Krasnozerskoye Goose
- Kuban Goose
- Leine Goose (*Leinegans*)
- Lippe Goose (*Lippegans*)
- Normandy Goose
- Norwegian White Goose
- Obroshin Goose
- Öland Goose
- Pereyaslavl Goose
- Pilgrim Goose
- Pomeranian Goose (including Pomeranian Saddleback)
- Pskov Goose
- Smålen Goose
- Rhineland Laying Goose (*Rheinische Legegans*, extinct breed)
- Roman Tufted Goose
- Russian Goose
- Scania Goose
- Sebastopol Goose (*Lockengans*)
- Shetland Goose
- Slovak White Goose (*Slovenská biela hus*)
- Steinbach Fighting Goose (*Steinbacher Kampfgans*)
- Subcarpathian Landrace (*Podkarpacka*)
- Suwa³ki Landrace (*Suwa³ska*)

- Suchovy Goose (*Suchovská hus*)
- Toulouse Goose (including Light Toulouse)
- Touraine Goose
- Tula Fighting Goose
- Twente Landrace (*Twentse landgans*)
- Ural Goose or Shadrin Goose
- Viðtinës Goose
- Vladimir Goose
- West of England Goose
- Zatory Landrace (*Zatorska*).

Auto-sexing Goose

The plumage of male and female goose is usually the same. However, there are few auto-sexing goose, which are sexually dimorphic and the gender can be recognised on the first look by plumage. In general, ganders are white and females are either entirely gray, or pied gray and white.

- Cotton Patch Goose
- Normandy Goose
- Pilgrim Goose
- Shetland Goose
- West of England Goose.

Semi-domesticated goose species:

- Canada Goose
- Egyptian Goose.

List of Turkey Breeds

There are currently eight breeds of domestic turkeys recognised by the American Poultry Association, but many more exist as officially unrecognised variants or as recognised breeds in other countries.

- Bronze Turkey, the heritage strain of the Bronze is recognised, while the Broad Breasted Bronze, like the Broad Breasted white, is an unrecognised commercial meat strain.
- Narragansett Turkey
- White Holland Turkey
- Black Spanish Turkey (Black Norfolk Turkey)
- Blue Slate Turkey

- Bourbon Red Turkey
- Beltsville Small White Turkey
- Royal Palm Turkey
- Auburn, an extremely rare heritage breed, numbers are not considered high enough for inclusion in the Standard.
- Broad Breasted White, a non-standardised commercial strain that does not qualify as a breed, only used for commercial meat production.
- Broad Breasted Bronze, a non-standardised commercial strain that does not qualify as a breed, only used for commercial meat production.
- Buff (or Jersey Buff)
- Chocolate, Chocolate Brown in colour. Day-old poults are white faced with chocolate bodies.
- Midget White, a rare heritage breed sometimes conflated with the Beltsville Small White.

Chapter 7

General Anatomy of Poultry

4-H Poultry Activity Guide

As you review the 4-H Activity Guide the following appendix may serve as a reference for your convenience.

To meaningfully study and recognise the distinguishable characteristics of the different species, breeds, and varieties of poultry, it will be necessary to know the accepted nomenclature of external anatomical features.

The male and female chickens have some identical features. It is desirable to recognise and distinguish the features of the beak, comb, ears, earlobes, eyes, eye ring (eyelid), hackles, thigh, lower leg, hock joint (ankle), shank (foot), toes, and claw. The lower part of the beak is hinged at the jaw and is movable; the upper part of the beak is fused to the skull.

The comb and wattles are red, soft, and warm. The ears are merely openings into the auditory canal protected by small feathers; the earlobes consist of tightly fitting specialised skin devoid of feathers. The colour of the earlobes (red or white) depends upon the breed. The eyeball is covered by the eye ring which, when open, appears as a circle of skin defining the ocular opening. The hackles are the feathers of the neck. The thighs are not easily distinguished in the standing chicken as they are located along each side of the body and well covered with feathers. The lower leg is feathered and articulates at the hock joint with the scaly shank. Since the chicken stands and walks on its toes, the shank is the foot and the hock joint is the ankle. Most chickens have three toes projecting forward and one claw projecting back.

There are different types of combs that are inherited characteristics of breeds and varieties. The single comb is most familiar, having its base of attachment to the skull. Its posterior edge is the blade, and the spaces defined by its points are serrations. The pea comb has three rows of bumps. The rose comb has many very small bumps and may not have a spike projecting back. The strawberry comb has a pitted texture, is relatively small, and sets well forward on the head with its larger end forward.

The v-shaped comb is associated with chickens that have a crest of feathers on the head, is very small, and sets well forward on the head. These chickens may or may not also have muffs or a beard of feathers. The buttercup comb, starting at the base of the beak forms a cup-shaped circle of points defining a deep cavity. It has a smooth, fine texture. The cushion comb is relatively small and smooth in texture, setting low and well forward on the head.

The observable differences in secondary sex characteristics between the male and female chicken are referred to as sexual dimorphism. The male has a larger body, comb, and wattles. In single-comb birds the male's comb will be turgid and stand erect, whereas the female's comb may flop over on one side. In multicoloured varieties, the male will have more variety of colouring in his plumage than the female. The male has longer and more pointed hackle feathers than the female. The male and female both have main tail feathers. However, in the male only, the tail feathers are covered by sickle feathers. Also, only the male has saddle feathers. The male has a larger, more developed spur than does the female. A young chicken from hatch to five weeks of age is called a chick. A male chicken less than one year of age is a cockerel; a female through her first laying year may be referred to as a pullet. A mature male chicken greater than one year of age is referred to as a cock or rooster; a mature female greater than one year old may be called a hen.

The turkey has nomenclature similar to the chicken but with a few notable differences. It has no comb on its head, but does have a fleshy growth from the base of its beak that is known as a snood, which is very long on males and hangs down over the beak. It has a wattle, but also bumpy, red, fleshy tissue covering the head and neck called caruncles. Male turkeys have a tuft of long, bristly, black, coarse fibres attached to the breast, known as the beard.

A young turkey is called poult. A male turkey of any age may be referred to as tom; female turkey, a hen. Ducks have nomenclature similar to that of the chicken, with the following notable differences.

There is no comb or other head covering. The duck's bill is flatter than the chicken's beak and has a protrusion on the upper tip known as a bean. The duck has webbed toes used for swimming. Male ducks have curled feathers at the base of the tail distinguishing them from females. Male ducks emit only a hiss, whereas the female will also emit a squawk when handled.

A young duck is called a duckling. An adult male is a drake; and an adult female, a duck. Geese have a few additional distinguishing features. Some breeds will have a horny knob at the base of the bill. Some geese also have dewlap, which is a loosely suspended growth of skin extending from the base of the lower bill along the upper throat. A young goose is called a gosling. An adult male is a gander; and an adult female, a goose. Pigeons, guineas, and various ornamental and game birds are frequently raised for pleasure. Also, a limited number of producers raise them for profit, on a full-time or part-time basis. Game birds are Bill raised for sale to game preserves or for shooting preserves. Also, there is a limited market for the sale of ornamental birds.

The domestic guinea fowl is descended from one of the wild species of Africa. Guineas might be more popular were it not for their harsh and seemingly never-ending cry, and their bad disposition. Guinea chicks are known as baby keets. Usually, sex can be distinguished by the cry and by the larger helmet and wattles and coarser head of the male.

The peafowl belongs to the same family as pheasants and chickens, differing in no important characteristic other than plumage. Peafowl have a very raucous voice, which may annoy neighbours. Pheasants are similar to chickens structurally and may be produced in a similar manner. Pheasants are generally raised for the purpose of stocking farms reserved for hunting by sportsmen. Pheasants originated in the orient and were first brought to America by Benjamin Franklin's son-in-law. Pheasants are classified as (1) game breeds, or (2) ornamental breeds.

Pigeons are a versatile bird with four distinct uses: (1) the sport of racing pigeons; (2) flyers and performers; (3) showing fancy pigeons; and (4) meat production. There are about 200 different breeds of pigeons, each distinct from the other in behaviour, size, shape, stance, feather form, colours, markings, and ornamentation. Pigeons are the most rapid growing of all poultry. Swans are an ornamental bird. Swan chicks are properly called cygnets. Swans respond to the same care as geese. Swans live to be very old; the males have been known to live for more than 60 years.

Eggs

Eggs are a biological structure intended by nature for reproduction of birds. They protect the developing chick embryo and provide food for the first few days of the chick's life. The egg is also one of the most nutritious and versatile of human foods.

Eggs of domestic chickens may be white, many shades of brown, or yellow. One breed lays bluegreen eggs. Sometimes very small, dark flecks are present on the eggshell, especially if it is brown. Egg colour often assumes economic importance, as there are numerous local prejudices in favour of shell tints. Coloured eggs occur because pigment is deposited in the shell as it is formed in the uterus.

The protective covering known as the shell is composed primarily of calcium carbonate, with 6,000 to 8,000 microscopic pores permitting transfer of volatile compounds. The air cell is located in the large end of the egg, and is formed when the cooling egg contracts and pulls the inner and outer shell membranes apart.

The chordlike chalazae holds the yolk in position in the centre of the egg. The germinal disc, a normal part of every egg, is located on the surface of the yolk. Embryo formation begins here only in fertilised eggs.

Parts of the Egg

The albumen, or egg white, is secreted around the yolk. Four distinct layers of albumen can be recognised in an egg: the chalaziferous layer, attached to the yolk; the inner thin albumen; the thick albumen; and the outer thin albumen. Three-fourths of the albumen is made up of the thick and outer thin albumen.

The twisting of the egg during formation appears responsible for the separation of the albumen into the four layers. Two shell membranes are formed, an inner and an outer shell membrane. These are rather loose fitting membranes when first formed. Water is added to the egg to "plump out" the egg into its final shape. The outer shell membrane is about three times as thick as the inner membrane. The membranes normally adhere to each other except at the large end of the egg, where they are separated to form the air cell.

The eggshell is made up almost entirely of calcium carbonate deposited on the outer shell membrane. The process of forming the shell requires 19 to 20 hours. About two grams of calcium is deposited in each eggshell. Strong eggshells are essential for eggs to be handled as they progress from farm to market. Hens are usually fed a laying ration to obtain the majority of the eggshell calcium directly from the feed, but they also withdraw some calcium from their bones, especially at night when they are not eating.

Digestion

Any animal can be thought of as a biological "machine" that converts raw materials into a finished product: in the case of poultry feed into meat and eggs; in the case of humans, food into happy, healthy, productive world citizens. Feed will pass, in order, through the following parts of the birds digestive tract: mouth, esophagus, crop, lower esophagus, proventriculus (glandular stomach), gizzard (muscular stomach), small intestine, ceca, large intestine (rectum), and cloaca. Not all ingesta goes through the ceca, which are mainly for breakdown of dietary fibre.

A distinctive characteristic of birds is the absence of lips and teeth. Instead, the bird has a hard beak that can be used for grasping, tearing, and scooping food. The digestive system works very efficiently in handling various types of food materials. The tongue contains the hyoid bone hinge at the lower jaw, and is pointed at the anterior tip with several barbed points projecting posteriorly on each side. Since the bird cannot swallow, the tongue moves back and forth forcing food down the esophagus.

There are a few saliva glands in the mouth that contribute some moisture to the feed at this point.

The esophagus (gullet) is part of the tube that conveys feed from the mouth ticles move rapidly into the muscular stomach (gizzard) where physical breakdown starts. Gizzards are highly muscular organs used for grinding and mixing feed materials in preparation for digestion. Feed leaving the gizzard passes into the duodenal loop of the upper small intestine.

The liver produces bile that is temporarily stored in the gall bladder. From the gall bladder, bile mixes with the food slurry as it passes into the next part of the small intestine. In the duodenal loop digestion starts as the pancreas secretes digestive enzymes. In the remaining area of the small intestine the digestive process is completed and absorption of nutrients takes place. The small intestine in a mature chicken is over 4.5 feet in length and terminates at its juncture with the large intestine.

The large intestine is relatively short, only about 4 inches in length, terminating at the cloaca. The ceca consists of two pouches that fill and empty from the same direction. Their main function is associated with breakdown of fibre, although chickens and turkeys cannot utilise large amounts commonly associated with some poultry diets. The major functions of the large intestine are storage of undigested waste material and absorption of water from their content.

The cloaca is the common chamber into which the digestive, urinary, and reproductive tracts open. Its opening at the posterior end of the bird is known as the vent. When the bird eliminates faecal waste from its digestive tract, the cloaca actually folds back at the vent allowing the rectal opening of the large intestine to push out, closing the reproductive opening. Thus, there is minimal chance of faecal wastes contaminating the reproductive system.

An understanding of the structures and function of the digestive tract of the bird is important to understand the need for highly specialised diets: low in fibre and containing all the necessary nutrients in adequate amounts that are relatively easy to digest. Closely associated with the digestive system in the process of excretion in the urinary system, or excretory system, including the elimination of waste products of body metabolism.

The kidneys are paired; each consisting of three lobes dorsally located along the vertical column posterior to the lungs. The ureters are long tubes that connect the kidneys with the cloaca for the purpose of transporting the waste products out of the body. The bird has no urinary bladder and thus does not produce a watery urine, as do mammals; it excretes the urates or products of metabolism as solids that are added to feces as a white cap.

A chicken is a type of domesticated bird which is often raised as a type of poultry. It is believed to be descended from the wild Indian and southeast Asian Red Junglefowl.

With a population of more than 24 billion in 2003, there are more chickens in the world than any other bird. They provide two sources of food frequently consumed by humans: their meat, also known as chicken, and eggs.

Appearance

Feathers, Comb, and Wattle: In some breeds, such as the Seabright, the cock only has slightly pointed neck feathers, and the identification must be made by looking at the comb.

Chickens have a fleshy crest on their heads called a comb, and a fleshy piece of hanging skin under their beak called a wattle. These organs help to cool the bird by redirecting blood flow to the skin. Both the male and female have distinctive wattles and combs. In males, the combs are often more prominent, though this is not the case in all varieties.

Characteristics and Behaviour

Feeding Habits: Chickens are omnivores and will feed on small seeds, herbs and leaves, grubs, insects and even small mammals if they can get them.

Domestic chickens are typically fed commercially prepared feed that includes a protein source as well as grains. Chickens often scratch at the soil to get at adult insects and larvae or seed.

Incidents of cannibalism: can occur when a curious bird pecks at a preexisting wound or during fighting (even among female birds). This is exacerbated in close quarters. In commercial egg production this is controlled by trimming the beak (removal of 2/3 of the top half and occasionally 1/3 of the lower half of the beak).

Can Chickens Fly: Domestic chickens are not capable of flying for long distances, although they are generally capable of flying for short distances such as over fences.

Chickens will sometimes fly simply in order to explore their surroundings, but will especially fly in an attempt to flee when they perceive danger. Because of the risk of flight, chickens raised in the open air generally have one of their wings clipped by the breeder — the tips of the longest feathers on one of the wings are cut, resulting in unbalanced flight which the bird cannot sustain for more than a few metres (more on wing clipping).

Pecking Order: Chickens are gregarious birds and live together as a flock. They have a communal approach to the incubation of eggs and raising of young. Individual chickens in a flock will dominate others, establishing a "pecking order", with dominant individuals having priority for access to food and nesting locations. Removing hens or roosters from a flock causes a temporary disruption to this social order until a new pecking order is established.

Crowing: Contrary to popular belief, roosters may crow at any time of the day or night. Their crowing - a loud and sometimes shrill call - is a territorial signal to other roosters. However, crowing may also result from sudden disturbances within their surroundings.

Chickens as pets: Chickens can make loving and gentle companion animals. In Asia, chickens with striking plumage have long been kept for ornamental purposes, including feather-footed varieties such as the Cochin and Silkie from China and the extremely long-tailed Phoenix from Japan. Asian ornamental varieties were imported into the United States and Great Britain in the late 1800s. Poultry fanciers then began keeping these ornamental birds for exhibition, a practice that continues today. From these Asian breeds, distinctive American varieties of chickens have been developed.

Today, some cities in the United States still allow residents to keep live chickens as pets, although the practice is quickly disappearing. Individuals in rural communities commonly keep chickens for both ornamental and practical value. Some communities ban only roosters, allowing the quieter hens. Many zoos use chickens instead of insecticides to control insect populations.

Raising Your Chicken: Growing chickens can easily be tamed by feeding them a special treat such as meal worms in the palm of one's hand, and by being with them for at least ten minutes daily when they are young. Chickens are also fond of the taste of beer.

A former recurring skit on the weekly comedy show Saturday Night Live featured a chicken pet store with the Chinese owner (as played by Dana Carvey) not wishing to sell to customers on the basis that "Chickens make lousy house pets."

Reproduction and OFFspring

Courting: When a rooster finds food he may call the other chickens to eat it first. He does this by clucking in a high pitch as well as picking up and dropping the food. In some cases the rooster will drag the wing opposite the hen on the ground, while circling her. This is part of chicken courting ritual. When a hen is used to coming to his "call" the rooster may mount the hen and proceed with the fertilization.

Nests and Eggs: Chickens will try to lay in nests that already contain eggs, and have been known to move eggs from neighbouring nests into their own.

Some farmers use fake eggs made from plastic or stone to encourage hens to lay in a particular location. The result of this behaviour is that a flock will use only a few preferred locations, rather than having a different nest for every bird.

Hens can also be extremely stubborn about always laying in the same location. It is not unknown for two (or more) hens to try to share the same nest at the same time. If the nest is small, or one of the hens is particularly determined, this may result in chickens trying to lay on top of each other.

Going Broody: Sometimes a hen will stop laying and instead will focus on the incubation of eggs, a state that is commonly known as going broody. A broody chicken will sit fast on the nest, and protest or peck in defence if disturbed or removed, and will rarely leave the nest to eat, drink, or dust bathe. While broody, the hen keeps the eggs at a constant temperature and humidity, as well as turning the eggs regularly.

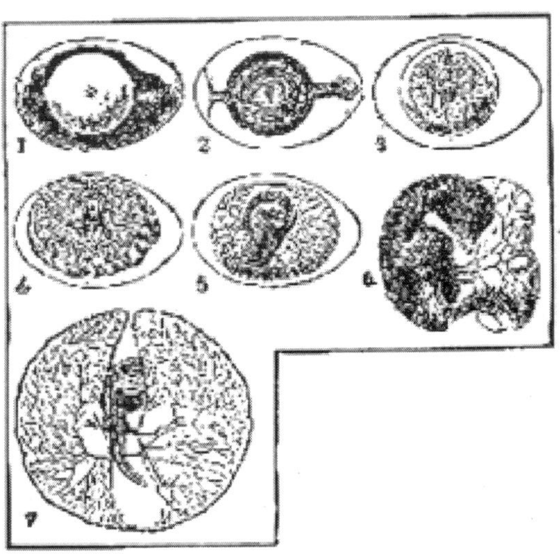

Stages of Chicken Embryo Growth

When Eggs Hatch: At the end of the incubation period, which is an average of 21 days, the eggs (if fertilised) will hatch, and the broody hen will take care of her young.

Since individual eggs do not all hatch at exactly the same time (the chicken can only lay one egg approximately every 25 hours), the hen will usually stay on the nest for about two days after the first egg hatches. During this time, the newly-hatched chicks live off the egg yolk they absorb just before hatching.

The hen can hear the chicks peeping inside the eggs, and will gently cluck to encourage them to break out of their shells. If the eggs are not fertilised and do not hatch, the hen will eventually grow tired of being broody and leave the nest.

Modern egg-laying breeds rarely go broody, and those that do often stop part-way through the incubation cycle. Some breeds, such as the Cochin, Cornish and Silkie, regularly go broody and make excellent mothers.

Artificial incubation: Chicken egg incubation can successfully occur artificially as well. Nearly all chicken eggs will hatch after 21 days of good conditions - 99.5° fahrenheit (37.5°C) and around 55% relative humidity (increase to 70% in the last three days of incubation to help soften egg shell).

Many commercial incubators are industrial-sized with shelves holding tens of thousands of eggs at a time, with rotation of the eggs a fully automated process.

Home incubators: are usually small boxes (styrofoam incubators are popular) and hold a few to 50 eggs. Eggs must be turned three to five times each day, rotating at least 90 degrees. If eggs aren't turned, the embryo inside will stick to the shell and likely will be hatched with physical defects. This process is natural; hens will stand up three to five times a day and shift the eggs around with their beak.

General Classification and Information

History of the Chicken: Chickens are domesticated descendants of the red junglefowl, which is biologically classified as the same species.

Red Jungle Fowl

Its In the Genes: Recent studies have shown that chickens (and possibly other bird species) still retain the genetic blueprints to produce teeth in the jaws, although these are dormant in living animals. These are a holdover from primitive birds such as Archaeopteryx, which were descended from theropod dinosaurs.

Atwater's Prairie Chicken

History of Chicken Sales: In the United States, chickens were once raised primarily on family farms. Prior to about 1930, chicken was served primarily on special occasions or on Sunday, as the birds were typically more valued for their eggs than meat. Excess roosters or nonproductive hens would be culled from the flock first for butchering. As cities developed and markets sprung up across the nation, live chickens from local farms could often be seen for sale in crates outside the market to be butchered and cleaned on site by the butcher.

Changes in Chicken Industry: With the advent of vertical integration and selective breeding of efficient meat-type birds, poultry production changed dramatically. Large farms and packing plants emerged that could grow birds by the thousands.

Chickens could be sent to slaughterhouses for butchering and processing into prepackaged commercial products to be frozen or shipped fresh to markets or wholesalers.

Meat-type chickens currently grow to market weight in 6-7 weeks whereas only fifty years ago it took three times as long. This is due exclusively to genetic selection and nutritional advances (and not to use of growth hormones, which are illegal for use in poultry in the US). Once a meat consumed only occasionally, the common availability has made chicken a common and significant meat product within developed nations. Growing concerns over the cholesterol content of red meat in the 1980s and 1990s further resulted in increased consumption of chicken.

Efficient Egg Layers: Another breed of chicken, the Leghorn, was further developed to be efficient layers of eggs. Egg production and consumption changed with the development of automation and refrigeration.

Large farms were devoted solely to egg production and packaging. Today, eggs are produced on large egg ranches on which environmental parametres are well controlled.

Chickens are exposed to artificial light cycles to stimulate egg production year-round. In addition, it is a common practice to induce moult through careful manipulation of light and the amount of food they receive in order to further increase egg size and production.

Issues with mass production: Many animal rights advocates object to killing chickens for food or to the "factory farm conditions" under which they are raised.

They contend that commercial chicken production often involves raising the birds in large, crowded rearing sheds that prevent the chickens from engaging in many of their natural behaviours. Contrary to popular belief, however, meat-type chickens are not raised in cages and are instead raised on the floor on litter such as rice hulls.

They are slaughtered prior to sexual maturity, and thus many of the aggressive behaviours seen in adult chickens (fighting, cannibalism) are seldom seen in meat-type chickens. In 2004, 8.9 billion chickens were slaughtered in the United States.

Bird Brains? Although many would argue that the birds are not intelligent and thus not a high priority for humane treatment on farms, a woman once brought a chicken on The Tonight Show with Jay Leno where it played "Mary Had A Little Lamb" on a toy piano and bowled 3 strikes. Animal rights groups such as PETA see this and other "amazing" trained chickens as evidence that they are intelligent and sentient and should not be killed or eaten.

Selective Breeding: Another animal welfare issue is the use of selective breeding to create heavy, large-breasted birds, which can lead to crippling leg disorders and heart failure for some of the birds.

In addition, many scientists have raised concerns that companies growing one variety of bird for eggs or meat are causing them to become much more susceptible to disease. For this reason, many scientists are promoting the conservation of heritage breeds to retain genetic diversity in the species.

Chicken Diseases: Chickens are susceptible to a wide variety of lethal diseases. Chickens are also susceptible to parasites, including lice, mites, ticks, fleas, and intestinal Worms.

Culture and Mythology: In Indonesia the chicken has great significance during the Hindu cremation ceremony.

A chicken is a channel for evil spirits which may be present during the ceremony. A chicken is tethered by the leg and kept present at the ceremony for the duration to ensure that any evil spirits present during the ceremony go into the chicken and not the family members present. The chicken is then taken home and returns to its normal life. It is not treated in any special way or slaughtered after the ceremony.

In ancient Greece, the chicken was not normally used for sacrifices, perhaps because it was still considered an exotic animal. Because of its valour, cocks are found as attributes of Ares, Heracles and Athena. The Greeks believed that even lions were afraid of cocks. Several of Aesop's Fables reference this belief. In the cult of Mithras, the cock was a symbol of the divine light and a guardian against evil.

In the Bible, Jesus prophesied the betrayal by Peter: "And he said, I tell thee, Peter, the cock shall not crow this day, before that thou shalt thrice deny that thou knowest me." (Luke 22:43) Thus it happened (Luke 22:61), and Peter cried bitterly.

This made the cock a symbol for both vigilance and betrayal. Earlier, Jesus compares himself to a mother hen, when talking about Jerusalem: "How often would I have gathered thy children together, even as a hen gathereth her chickens under her wings, and ye would not!" (Matthew 23:37; also Luke 13:34). In many Central European folk tales, the devil is believed to flee at the first crowing of a cock.

In some sects of Orthodox Judaism a chicken is slaughtered on the afternoon before Yom Kippur (Day of Atonement) in a ceremony called kappores.

Although not actually a sacrifice in the biblical sense, the death of the chicken reminds the penitent sinner that his or her life is in God's hands.

A woman brings a hen to be slaughtered, a man brings a rooster. The meat is donated to the poor.

The Talmud speaks of learning "courtesy toward one's mate" from the rooster. This might refer to the fact that, when a rooster finds something good to eat, he calls his hens to eat first.

Greater Prairie Chicken

The chicken is one of the Zodiac symbols of the Chinese calendar. Also in Chinese religion, a cooked chicken as a religious offering is usually limited to ancestor veneration and worship of village deities. Vegetarian deities such as Buddha are not one of the recipients of such offerings. Under some observations, an offering of chicken is present with "serious" prayer (while roasted pork is offered during a joyous celebration).

History: The first pictures of chickens in Europe are found on Corinthian pottery of the 7th century BC. The poet Cratinus (mid 5th century BC, according to the later Greek author Athenaeus) calls the chicken "the Persian alarm". In Aristophanes's comedy The Birds (414 BC) a chicken is called "the Median bird", which points to an introduction from the East. Pictures of chickens are found on Greek red figure and black-figure pottery.

In ancient Greece, chickens were still rare and were a rather prestigious food for symposia. Delos seems to have been a centre of chicken breeding. An early domestication of chickens in Southeast Asia is probable, since the word for domestic chicken (*manuk) is part of the reconstructed Proto-Austronesian language. Chickens, together with dogs and pigs, were the domestic animals of the Lapita culture, the first Neolithic culture of Oceania.

Spread of the Chicken: Chickens were spread by Polynesian seafarers and reached Easter Island in the 12th century AD, where they were the only domestic animal, with the possible exception of the

Polynesian Rat (Rattus exulans). They were housed in extremely solid chicken coops built from stone. Travelling as cargo on trading boats, they reached the Asian continent via the islands of Indonesia and from there spread west to Europe and western Asia.

Chickens in Ancient Rome: The Romans used chickens for oracles, both when flying and when feeding. The hen gave a favourable omen, when appearing from the left, like the crow and the owl. For the oracle, according to Cicero, any bird could be used, but normally only chickens were consulted. The chickens were cared for by the pullarius, who opened their cage and fed them pulses or a special kind of soft cake when an augury was needed. If the chickens stayed in their cage, made noises, beat their wings or flew away, the omen was bad; if they ate greedily, the omen was good.

Per Columella, the ideal flock consists of 200 birds, which can be supervised by one person if someone is watching for stray animals. White chickens should be avoided as they are not very fertile and are easily caught by eagles or goshawks. One cock should be kept for five hens. In the case of Rhodian and Median cocks that are very heavy and therefore not much inclined to sex, only three hens are kept per cock. The hens of heavy fowls are not much inclined to brood; therefore their eggs are best hatched by normal hens. A hen can hatch no more than 15-23 eggs, depending on the time of year, and supervise no more than 30 hatchelings. Eggs that are long and pointed give more male, rounded eggs mainly female hatchlings.

Turkeys

Turkeys are raised primarily for meat.

Consumers want birds that have a high proportion of white breast meat. The United States produces nearly 300 million turkeys each year.

Turkeys may vary in colour from white to bronze with mottled shades of black. The mottled shades are not as common as white or bronze.

White turkeys are the most popular turkeys for the production of meat. Others breeds can be bronze (red) or black coloured. This bird is strutting, fluffing its feathers.

Geese

About 1 million geese are raised in the United States each year. Geese are raised for meat, eggs, feathers and down. (Down is the soft feathery covering that grows under feathers.) Many geese are kept for ornamental purposes. Some geese are kept to control weeds and grass.

The White Chinese goose has a distinctive knob on its head. Chinese geese can be coloured brown in addition to the white colour. The Embden was one of the first breeds of geese introduced to the United States. It originated in Germany.

The Toulouse goose originates from the Toulouse area of southern France. The plumage is dark gray on the back, gradually shading to light gray edged with white on the breast and to white on the abdomen.

Ducks

Ducks are raised for meat, eggs, down and feathers. (Down is the soft feathery covering that grows under feathers.) Ducks are also kept as hobby or ornamental ducks.

White Pekin ducks are the most popular meat duck in the United States reaching a market weight at 7 pounds in 8 weeks. The breed originated in China and was brought to the United States in the 1870s. Ducks are very versatile and live happily under a wide variety of climatic conditions.

Virtually everything from feathers to feet, including the liver and tongue, can be turned into a profit; the only unusable thing about them is their quack. Rouen ducks are excellent meat producers but poor egg production and coloured plumage make them unsuitable for mass commercial production. White plumage is preferred for commercial feather processing. Muscovy ducks originated in South America. Numerous varieties of Muscovies exist; the white variety is the most desirable for market purposes. Mucsovies are an excellent meat bird but their low egg production makes them unsuitable for commercial duck farms. Although they are not ideally suited to commercial production, Muscovies have excellent possibilities for small general farms with special retail outlets. Call ducks are well suited as meat producers but poor egg production and coloured plumage make them unsuitable for commercial production.

Guinea Fowl

The brightly coloured plumage makes the *Guineas are raised for food, as novelty* birds, and to stock game preserves. Guinea Fowl get their name from Guinea, a part of the western coast of Africa. History reveals that Guinea fowl have been raised as table fare since before the time of the ancient Greeks and Romans. Guinea fowl are used as a substitute for game birds and are considered a delicacy in some restaurants. Guineas might be more popular in the United States if they were not so loud with their harsh and seemingly never ending cry. They often have a bad disposition and are not very popular with commercial producers.

Peafowl

Peafowl are raised for large, beautiful feathers. The feathers may be five times the length of the body.

Peafowl belongs to the same family as pheasants and chickens, differing in no important characteristic other than plumage. Peafowl are native to India, Burma, and Malaya.

Pigeons

Peafowl are usually sold as pairs of ornamental birds. They are edible and are *Pigeons are versatile with four distinct uses:* (1) the sport of racing pigeons; (2) flyers and performers; (3) showing fancy pigeons; and (4) meat production.

There are about 200 different breeds of pigeons, each distinct from the other in behaviour, size, shape, stance, feather form, colours, markings, and ornamentation. Pigeons often mate in pairs and remain pairs for life. Regarded as a delicacy for special occasions.

Chickens

Chickens primarily are raised for meat and eggs. The type raised depends on the product wanted. A few other speciality types are raised, such as game chickens and fancy show chickens.

Newly hatched chicks. Chicks is a term used to describe young chickens. Plymouth Barred Rock rooster. Roosters are the male of the species and hens are the female of the species. The Plymouth Rock is one of the foundation breeds of the modern broiler industry.

Plymouth Rock rooster has a single comb with red wattles. Generally, Plymouth Rocks are not extremely aggressive, and tame quite easily.

Rhode Island Reds are a good choice for the small flock owner. A dual-purpose medium heavy fowl; used more for egg production than meat production because of its darkcoloured pin feathers and its good rate of lay. Relatively hardy, they are probably the best egg layers of the dual-purpose breeds. This breed lays brown eggs.

New Hampshire a dual-purpose chicken, selected more for meat production than egg production. Medium heavy in weight, it dresses a nice, plump carcass as either a broiler or a roaster. This breed lays brown eggs.

White Leghorn is the most popular commercial egg production breed. Leghorns Buff Cochin hens do not have the elaborate combs and colouring of the roosters.

Old English game rooster. This breed is tightly feathered, very active, and very noisy.

Cochins are literally big, fluffy balls of feathers. They are mainly kept as an ornamental fowl and are well suited to close confinement. Their ability as mothers is widely recognised and Cochins are frequently used as foster mothers for game birds and other species. Cochins are originally from China but underwent considerable development in the U.S.

They lay brown eggs. Take their name from the city of Leghorn, Italy, where they are considered to have originated. Leghorns and their descendants are the most numerous breed we have in America today. This breed lays white eggs.

Bantams

Bantams are the miniatures of the poultry world. The word bantam is the overall term for more than 350 kinds of true-breeding miniature chickens. Black Breasted Red Old English Game Bantam roosters were once popular as fighting birds until the sport was outlawed. Today they are bred as ornamental birds. Black Breasted Red Old English Game Bantam hens are not as colourful as roosters but make attractive exhibits.

Black Bantam roosters are raised primarily as ornamental birds. Bantams are produced in a very large range of colour markings. Bantams will commonly have a name similar or like the standard chicken breeds followed by bantam.

Bird Anatomy

Bird anatomy, or the physiological structure of birds' bodies, shows many unique adaptations, mostly aiding flight.

Birds have a light skeletal system and light but powerful musculature which, along with circulatory and respiratory systems capable of very high metabolic rates and oxygen supply, permit the bird to fly.

The development of a beak has led to evolution of a specially adapted digestive system.

These anatomical specialisation have earned birds their own class in the vertebrate phylum.

Respiratory System

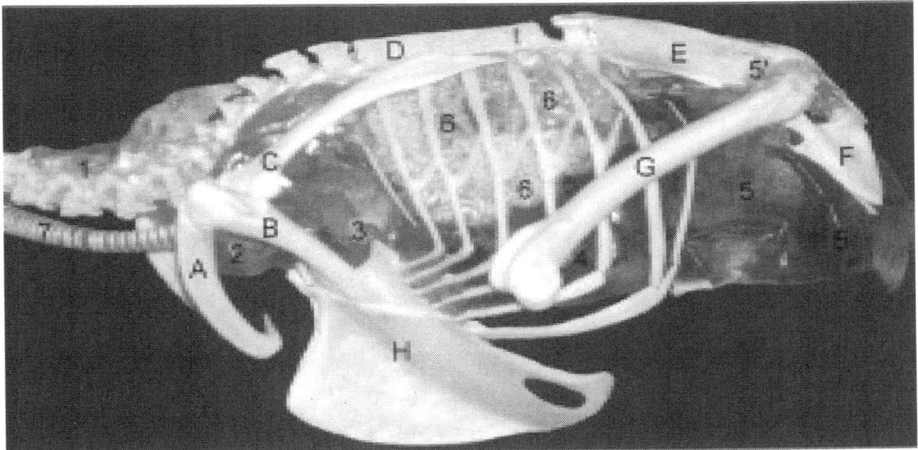

Air always flows from right (posterior) to left (anterior) through a bird's lungs during both inhalation and exhalation. Key to a Common Kestrel's circulatory lung system: 1 cervical air sac, 2 clavicular air sac, 3 cranial thoracic air sac, 4 caudal thoracic air sac, 5 abdominal air sac (5' diverticulus into pelvic girdle), 6 lung, 7 trachea

Due to their high metabolic rate required for flight, birds have a high oxygen demand. Development of an efficient respiratory system enabled the evolution of flight in birds. Birds ventilate their lungs by means of air sacs.

These sacs do not play a direct role in gas exchange, but to store air and act like bellows, allowing the lungs to maintain a fixed volume with fresh air constantly flowing through them.

Three distinct sets of organs perform respiration—the anterior air sacs (interclavicular, cervicals, and anterior thoracics), the lungs, and the posterior air sacs (posterior thoracics and abdominals). The posterior and anterior air sacs, typically nine, expand during inhalation. Air enters the bird via the trachea. Half of the inhaled air enters the posterior air sacs, the other half passes through the lungs and into the anterior air sacs. Air from the anterior air sacs empties directly into the trachea and out the bird's mouth or nares.

The posterior air sacs empty their air into the lungs. Air passing through the lungs as the bird exhales is expelled via the trachea. Some taxonomic groups (Passeriformes) possess 7 air sacs, as the clavicular air sacs may interconnect or be fused with the cranial thoracic air sacs.

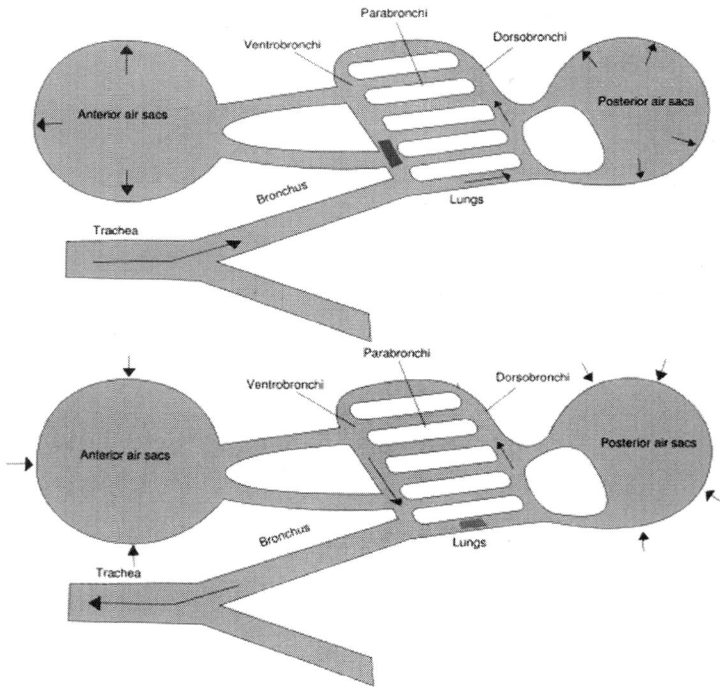

Figure : *Birds lungs obtain fresh air during both exhalation and inhalation*

As air flows through the air sac system and lungs, there is no mixing of oxygen-rich air and oxygen-poor, carbon dioxide-rich, air as in mammalian lungs. Thus, the partial pressure of oxygen in a bird's lungs is the same as the environment, and so birds have more efficient gas-exchange of both oxygen and carbon dioxide than do mammals. In addition, air passes through the lungs in both exhalation and inspiration, with the air sacs functioning as a reservoir for the next breath of air. Avian lungs do not have alveoli, as mammalian lungs do, but instead contain millions of tiny passages known as parabronchi, connected at either ends by the dorsobronchi and ventrobronchi. Air flows through the honeycombed walls of the parabronchi into air vesicles, called atria, which project radially from the parabronchi. These atria give rise to air capillaries, where oxygen and carbon dioxide are traded with cross-flowing blood capillaries by diffusion.

Birds also lack a diaphragm. The entire body cavity acts as a bellows to move air through the lungs. The active phase of respiration in birds is exhalation, requiring muscular contraction.

The syrinx is the sound-producing vocal organ of birds, located at the base of a bird's trachea. As with the mammalian larynx, sound is produced by the vibration of air flowing through the organ. The syrinx enables some species of birds to produce extremely complex vocalisations, even mimicking human speech. In some songbirds, the syrinx can produce more than one sound at a time.

Digestive System

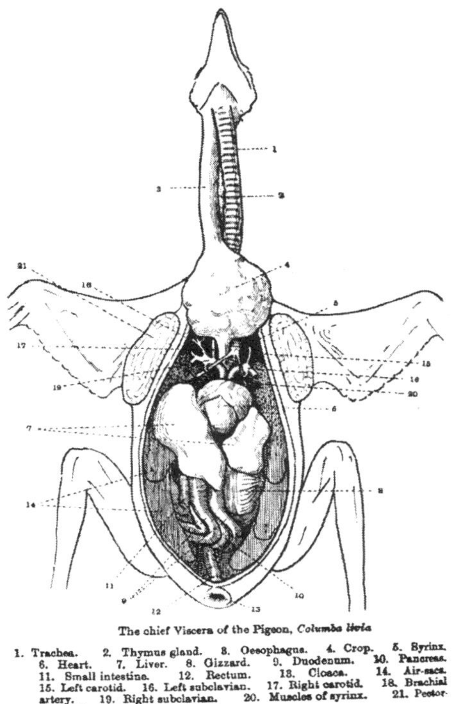

Figure : *Alimentary canal of the bird exposed.*

Figure : *Sharp tooth-like structures in this rooster's mouth called papillae help birds hold and move food around*

Many birds possess a muscular pouch along the esophagus called a crop. The crop functions to both soften food and regulate its flow through the system by storing it temporarily.

The size and shape of the crop is quite variable among the birds. Members of the order Columbiformes, such as pigeons, produce a nutritious crop milk which is fed to their young by regurgitation. Birds possess a *ventriculus*, or gizzard, composed of four muscular bands that rotate and crush food by shifting the food from one area to the next within the gizzard. The gizzard of some species contains small pieces of grit or stone swallowed by the bird to aid in the grinding process of digestion, serving the function of mammalian or reptilian teeth. The use of gizzard stones is a similarity between birds and dinosaurs, which left gizzard stones called gastroliths as trace fossils.

Circulatory System

Birds have a four-chambered heart, in common with humans, most mammals, and some other reptiles (namely the crocodilia). This adaptation allows for an efficient nutrient and oxygen transport throughout the body, providing birds with energy to fly and maintain high levels of activity. A Ruby-throated Hummingbird's heart beats up to 1200 times per minute (about 20 beats per second).

Drinking Behaviour

There are four general ways in which birds drink. Most birds are unable to swallow by the "sucking" or "pumping" action of peristalsis in their esophagus (as humans do), and drink by repeatedly raising their heads after filling their mouths to allow the liquid to flow by gravity, a method usually described as "sipping" or "tipping up". The notable exception is the Columbiformes; in fact, according to Konrad Lorenz in 1939, "one recognises the order by the single behavioural characteristic, namely that in drinking the water is pumped up by peristalsis of the esophagus which occurs without exception within the order. The only other group, however, which shows the same behaviour, the Pteroclidae, is placed near the doves just by this doubtlessly very old characteristic."

Although this general rule still stands, since that time, observations have been made of a few exceptions in both directions., In addition, specialised nectar feeders like sunbirds (Nectariniidae) and hummingbirds (Trochilidae) drink by using protrusible grooved or trough-like tongues, and parrots (Psittacidae) lap up water. Many seabirds have glands near the eyes that allow them to drink seawater. Excess salt is eliminated from the nostrils. Many desert birds get the water that they need entirely from their food. The elimination of nitrogenous wastes as uric acid reduces the physiological demand for water.

Skeletal System

The bird skeleton is highly adapted for flight. It is extremely lightweight but strong enough to withstand the stresses of taking off, flying, and landing. One key adaptation is the fusing of bones into single ossifications, such as the pygostyle. Because of this, birds usually have a smaller number of bones than other terrestrial vertebrates. Birds also lack teeth or even a true jaw, instead having evolved a beak, which is far more lightweight. The beaks of many baby birds have a projection called an egg tooth, which facilitates their exit from the amniotic egg.

Birds have many bones that are hollow (pneumatised) with crisscrossing struts or trusses for structural strength. The number of hollow bones varies among species, though large gliding and soaring birds tend to have the most. Respiratory air sacs often form air pockets within the semi-hollow bones of the bird's skeleton. Some flightless birds like penguins and ostriches have only solid bones, further evidencing the link between flight and the adaptation of hollow bones.

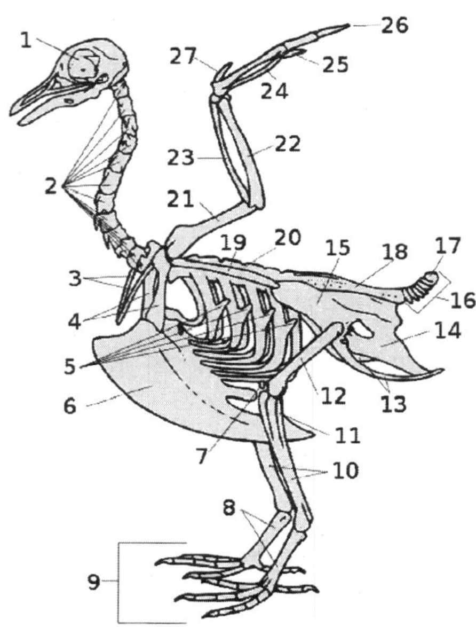

A stylised dove skeleton. Key:

1. skull
2. cervical vertebrae
3. furcula
4. coracoid
5. uncinate processes of ribs
6. keel
7. patella
8. tarsometatarsus
9. digits
10. tibia (tibiotarsus)
11. fibia (tibiotarsus)
12. femur
13. ischium (innominate)
14. pubis (innominate)
15. illium (innominate)
16. caudal vertebrae
17. pygostyle
18. synsacrum
19. scapula
20. lumbar vertebrae
21. humerus
22. ulna
23. radius
24. carpus
25. metacarpus
26. digits
27. alula.

Birds also have more cervical (neck) vertebrae than many other animals; most have a highly flexible neck consisting of 13-25 vertebrae. Birds are the only vertebrate animals to have a fused collarbone (the furcula or wishbone) or a keeled sternum or breastbone. The keel of the sternum serves as an attachment site for the muscles used for flight, or similarly for swimming in penguins. Again, flightless birds, such as ostriches, which do not have highly developed pectoral muscles, lack a pronounced keel on the sternum. It is noted that swimming birds have a wide sternum, while walking birds had a long or high sternum while flying birds have the width and height nearly equal.

Birds have uncinate processes on the ribs. These are hooked extensions of bone which help to strengthen the rib cage by overlapping with the rib behind them. This feature is also found in the tuatara *Sphenodon*. They also have a greatly elongate tetradiate pelvis as in some reptiles. The hindlimb has an intra-tarsal joint found also in some reptiles. There is extensive fusion of the trunk vertebrae as well as fusion with the pectoral girdle. They have a diapsid skull as in reptiles with a pre-lachrymal fossa (present in some reptiles). The skull has a single occipital condyle.

Skeleton

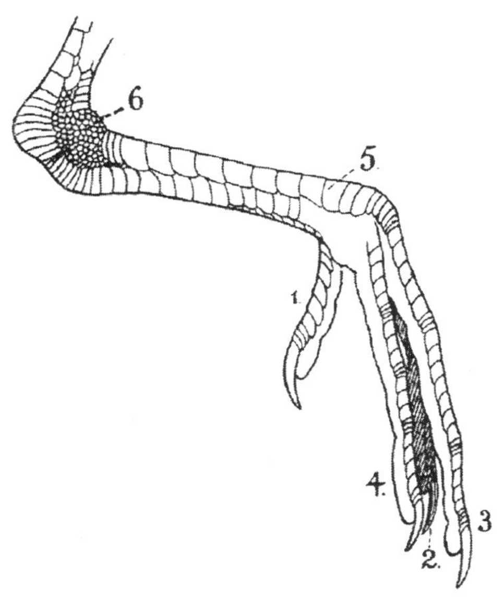

Side view of Right Foot of a Purple Gallinule (*Porphyrio*) to show the composition of the horny covering (*podotheca*).

1. Hallux or hind toe.
2. Inner toe.
3. Middle toe.
4. Outer toe.
5. Scales (*Scutellæ*).
6. Reticulate scales.

Figure : *Scalation and structure of the leg*

The skull consists of five major bones: the frontal (top of head), parietal (back of head), premaxillary and nasal (top beak), and the mandible. The skull of a normal bird usually weighs about 1% of the birds total bodyweight.

The vertebral column consists of vertebrae, and is divided into three sections: cervical (11-25) (neck), Synsacrum (fused vertebrae of the back, also fused to the hips (pelvis)), and pygostyle (tail). The chest consists of the furcula (wishbone) and coracoid (collar bone), which two bones, together with the scapula, form the pectoral girdle. The side of the chest is formed by the ribs, which meet at the sternum (midline of the chest).

The shoulder consists of the scapula (shoulder blade), coracoid, and humerus (upper arm). The humerus joins the radius and ulna (forearm) to form the elbow. The carpus and metacarpus form the "wrist" and "hand" of the bird, and the digits (fingers) are fused together. The bones in the wing are extremely light so that the bird can fly more easily.

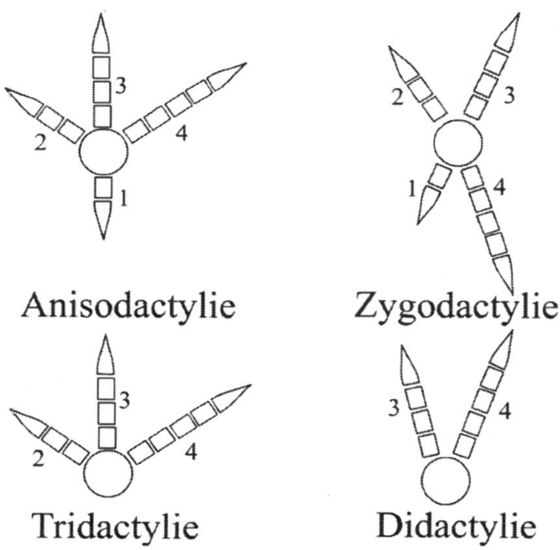

***Figure** : Types of bird feet*

The hips consist of the pelvis which includes three major bones: Illium, Ischium (sides of hip), and Pubis (front of the hip). These are fused into one (the innominate bone). Innominate bones are evolutionary significant in that they allow birds to lay eggs. They meet at the acetabulum (the hip socket) and articulate with the femur, which is the first bone of the hind limb. The upper leg consists of the femur. At the knee joint, the femur connects to the tibiotarsus (shin) and fibula (side of lower leg). The tarsometatarsus forms the upper part of the foot, digits make up the toes. The leg bones of birds are the heaviest, contributing to a low centre of gravity. This aids in flight. A bird's skeleton comprises only about 5% of its total body weight .

Birds feet are classificated as anisodactyl, zygodactyl, heterodactyl, syndactyl or pamprodactyl.

Muscular System

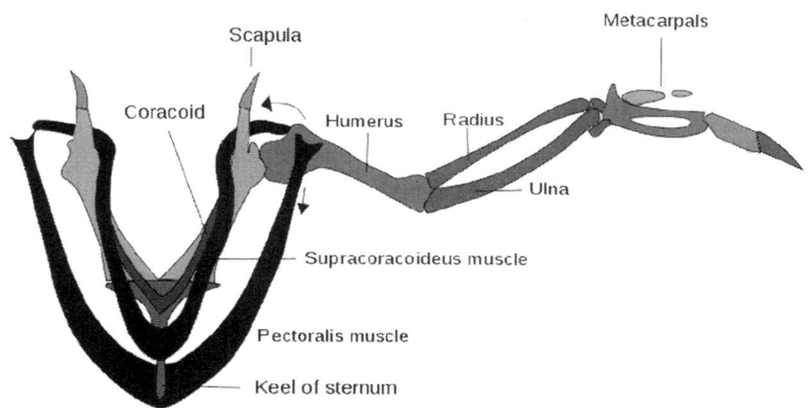

Figure: *The supracoracoideus works using a pulley like system to lift the wing while the pectorals provide the powerful downstroke*

Most birds have approximately 175 different muscles, mainly controlling the wings, skin, and legs. The largest muscles in the bird are the pectorals, or the breast muscles, which control the wings and make up about 15 - 25% of a flighted bird's body weight. They provide the powerful wing stroke essential for flight. The muscle medial (underneath) to the pectorals is the supracoracoideus. It raises the wing between wingbeats. The supracoracoideus and the pectorals together make up about 25 – 35% of the bird's full body weight.

The skin muscles help a bird in its flight by adjusting the feathers, which are attached to the skin muscle and help the bird in its flight maneuvers.

There are only a few muscles in the trunk and the tail, but they are very strong and are essential for the bird. The pygostyle controls all the movement in the tail and controls the feathers in the tail. This gives the tail a larger surface area which helps keep the bird in the air.

Head

Birds have acute eyesight - raptors have vision eight times sharper than humans - thanks to higher densities of photoreceptors in the retina (up to 1,000,000 per square mm in *Buteos*, compared to 200,000 for humans), a high number of optic nerves, a second set of eye muscles not found in other animals, and, in some cases, an indented fovea which magnifies the central part of the visual field. Many species, including hummingbirds and albatrosses, have two foveas in each eye. Many birds can detect polarised light. The eye occupies a considerable part of the skull and is surrounded by a sclerotic eye-ring, a ring of tiny bones that surround the eye. This character is also seen in the reptiles.

The bills of many waders have Herbst corpuscles which help them detect prey hidden under wet sand using minute pressure differences in the water. All extant birds can move the parts of the upper jaw relative to the brain case. However this is more prominent in some birds and can be readily detected in parrots.

Birds have a large brain to body mass ratio. This is reflected in the advanced and complex bird intelligence.

The region between the eye and bill on the side of a bird's head is called the lore. This region is sometimes featherless, and the skin may be tinted, as in many species of the cormorant family.

Reproduction

Although most male birds have no external sex organs, the male does have two testes which become hundreds of times larger during the breeding season to produce sperm. The testes in male birds are generally asymmetric with most birds having a larger left testis. Female birds in most families have only one functional ovary, connected to an oviduct - although two ovaries are present in the embryonic stage of each female bird. Some species of birds have two functional ovaries, and the order Apterygiformes always retain both ovaries.

In the males of species without a phallus, sperm is stored in the semenal glomera within the cloacal protuberance prior to copulation. During copulation, the female moves her tail to the side and the male either mounts the female from behind or in front (in the stitchbird), or moves very close to her. The cloacae then touch, so that the sperm can enter the female's reproductive tract. This can happen very fast, sometimes in less than half a second.

The sperm is stored in the female's sperm storage tubules for a week to more than a 100 days, depending on the species. Then, eggs will be fertilised individually as they leave the ovaries, before being laid by the female. The eggs continue their development outside the female body.

Many waterfowl and some other birds, such as the ostrich and turkey, possess a phallus. The length is thought to be related to sperm competition. When not copulating, it is hidden within the proctodeum compartment within the cloaca, just inside the vent.

After the eggs hatch, parents provide varying degrees of care in terms of food and protection. Precocial birds can care for themselves independently within minutes of hatching; altricial hatchlings are helpless, blind, and naked, and require extended parental care. The chicks of many ground-nesting birds such as partridges and waders are often able to run virtually immediately after hatching; such birds are referred to as nidifugous. The young of hole nesters, on the other hand, are often totally incapable of unassisted survival. The process whereby a chick acquires feathers until it can fly is called "fledging". Some birds, such as pigeons, geese, and Red-crowned Cranes, remain with their mates for life and may produce offspring on a regular basis.

Scales

The scales of birds are composed of the same keratin as beaks, claws, and spurs. They are found mainly on the toes and metatarsus, but may be found further up on the ankle in some birds. Most bird scales do not overlap significantly, except in the cases of kingfishers and woodpeckers. The scales and scutes of birds are thought to be homologous to those of reptiles and mammals.

Bird embryos begin development with smooth skin. On the feet, the corneum, or outermost layer, of this skin may keratinise, thicken and form scales. These scales can be organised into;

1. Cancella – minute scales which are really just a thickening and hardening of the skin, crisscrossed with shallow grooves.
2. Reticula – small but distinct, separate, scales. Found on the lateral and medial surfaces (sides) of the chicken metatarsus. These are made up of alpha-keratin.
3. Scutella – scales that are not quite as large as scutes, such as those found on the caudal, or hind part, of the chicken metatarsus.
4. Scutes – the largest scales, usually on the anterior surface of the metatarsus and dorsal surface of the toes. These are made up of beta-keratin as in reptilian scales.

The rows of scutes on the anterior of the metatarsus can be called an acrometatarsium or acrotarsium.

Feathers can be intermixed with scales on some birds' feet. Feather follicles can lie between scales or even directly beneath them, in the deeper dermis layer of the skin. In this last case, feathers may emerge directly through scales, and be encircled at the plane of emergence entirely by the keratin of the scale.

Chapter 8

Community-based Control Strategy

A comprehensive survey of the backyard poultry production system and its socioeconomic context was undertaken across 5 Provinces of Zimbabwe (Madzima *et al*, 1998a). The survey included a longitudinal production study, a seasonal study and informal interviews with women producers. The following discussion summarises key factors drawn from this survey relevant to the development of a community-based Newcastle disease control strategy.

Use of feed Grains as Vaccine Carriers

From trials into potential feed carriers for V4 and I2, only rapoko was found to be suitable, but not maize, rice, millet or sorghum. This indicates that the V4 and I2 vaccines are highly sensitive to the grain they are delivered on, and that no single grain type can be identified as suitable. This is confirmed by similar findings in both Africa and Asia. To develop a feed-borne vaccine strategy, therefore, requires examination of all potential grains in any country, and possible separate trials in different areas of the country, since grain characteristics may vary from region to region. There is also limited understanding of the impact on the vaccine of grain variety, quality and condition, chemical additives present, and also the surfaces grains are fed on.

Significantly, in Zimbabwe only 15 % of producers provide supplementary feed for their birds, and feeding regimes vary significantly. If feed grains are not a traditional input for many backyard producers, basing a vaccine strategy on their use represents a significant constraint. As an example, rapoko is not grown in all areas of the country, nor is it widely available at all times of the year (Mavhenyengwa, pers. comm.).

Although called 'heat-stable', no quantified definition has been applied to this term, and it is unclear how many hours and at what temperatures the vaccine can be kept before its potency deteriorates. Vaccine performance is also affected by excessive vibration and extended exposure to direct sunlight. Feed-based vaccination, however, inevitably requires vibration during mixing and exposure to sunlight during application. Finally, the feeding rates of individual birds are difficult to monitor and control. To achieve protection, individual birds must consume a given quantity of grain and vaccine, but feeding rates and pecking order in flocks can disadvantage younger and weaker birds.

Backyard Flock Dynamics

The epidemiology of Newcastle disease under backyard systems remains unclear (Awan *et al*, 1994), but it is estimated that within a population, the disease can be considered under control if less than 30% of birds are infected (Palya, pers. comm.). The average backyard flock in Zimbabwe numbers 20 birds, and is composed of 8 chicks, 6-7 growers, 4-5 hens and 1 cock (Oakeley, 1998a). Based on rates of egg incubation and loss, and mortality in chicks and growers, flock turn-over rate due to the introduction of 'new' birds, means that from a point in time, the average flock may comprise of 30% unprotected birds within 4 months. This indicates the need for vaccination between 2 and 3 times a year if effective cover is to be maintained in these flocks (Oakeley, 1998b).

These statistics must be viewed in the light of vaccine trial results. Field trials focused on the use of rapoko as a carrier for both V4 and I2, but also examined water-based application of V4, and eye-drop delivery of I2 and the conventional vaccine La Sota. Birds in the various trial groups were vaccinated every month, for up to 6 months, and tested for antibody levels using the haemagglutination inhibition test. (Madzima *et al*, 1998a). Both I2 and La Sota delivered by the intra-ocular route afforded adequate protection rates after a single application. Water-borne V4 achieved less than 40% protection of flocks after three vaccinations, and feed-borne I2 and V4 less than 30% and zero protection, respectively, after six applications (Palya, 1998).

These test results confirm findings from trials elsewhere in Africa and Asia. The findings suggest the need for further development and improvement of heat-stable vaccines before feed - or water-based delivery strategies can offer an effective technical solution to Newcastle disease control. However, there remains the question of how any control strategy can incorporate community participation, and thereby offer a more sustainable approach to Newcastle disease control in backyard poultry systems.

The Prioritisation of Newcastle Disease Control

Ultimately, maintaining control of Newcastle disease in backyard flocks will depend on the involvement of the farming community, and farmer enthusiasm reflects the level of priority they give the disease. A complex array of production and health constraints is associated with varying levels of technical, management and husbandry input in Zimbabwe. The survey identified massive variation between flocks in the levels of egg and bird off-take at every stage of the production cycle. While impossible to attribute to any single factor, this variation results from differential levels of egg and bird management, feeding, health care, chick rearing and other husbandry activities.

Despite the evident variation in backyard flock productivity, 78% of extensive producers surveyed claimed to have no support or contact with the formal veterinary and extension services. There is evidence that the research and extension system does not cater adequately, either for women producers, who are the primary stakeholders in extensive poultry, or for the system of extensive poultry production as a whole.

Poultry production must be seen as only one of many household and farm activities, and is rarely the priority concern, even of women who tend to value it more highly than men. Other responsibilities such as household duties, other livestock or seasonal crop-related work can take precedence over poultry activities. It is, therefore, not clear how much additional time some households are prepared to commit to activities like poultry disease control.

Customary practices can also complicate vaccination campaigns. The transfer and movement of chickens between villages and regions is a fundamental part of the extensive system, enabling owners to use birds for celebrations, gifts and as ready sources of cash. These movements influence the epidemiology of the disease, and complicate the monitoring and control of vaccination cover. Mass vaccination campaigns have also been hampered by absence of producers at key times, and reports of outbreaks following previous campaigns (Mavhenyengwa, pers. comm.).

Most producers appear well informed about Newcastle disease and its impact, but it is only one of numerous health constraints threatening backyard flocks. There is widespread incidence of fowl pox, infectious bursal disease, losses to predators, and internal and external parasites, as well as considerable management and husbandry constraints (Oakeley, 1998b). No information is available on what proportions of losses are a result of different diseases or problems, and no quantitative data are available on the specific losses attributable to Newcastle disease.

The absence of a complete production and health extension package specifically aimed at extensive poultry producers was highlighted by the study. Since a sustained control strategy depends upon producer commitment, the benefits of that commitment must be clear and acceptable to the farming community. While there remain gaps in the services offered to backyard producers, limited commitment toward isolated health campaigns such as Newcastle disease vaccination can only be expected. This may be one reason why extensive poultry producers have limited interest in Newcastle disease control.

In addition, it is not clear how great a threat backyard producers perceive New-castle disease to be. In commercial systems, the threat is significant when large numbers of birds, in confined areas, can die within short time periods as a result of an outbreak. In extensive flocks, however, some birds can be salvaged in the face of an outbreak, through consumption, sale or gift (Spradbrow, 1994). Such strategies will greatly effect the true economic loss associated with an outbreak, but may not be reflected in the statistics of birds that die or are slaughtered in response to an outbreak.

Concluding Discussion

A question remains over the readiness of backyard producers to adopt any strategy based on the use of supplementary feeds that do not currently feature in many backyard production systems. Not only does the purchase of feed grain represent a constraint to some producers, delivery procedures are also open to misapplication and problems of monitoring. Involving producers in feed-based vaccine delivery may, in practice, prove no less problematic than facilitating their participation in more conventional techniques. Zimbabwe now uses V4 delivered by intra-ocular route (Oakeley, 1998b), and is training individual producers how to handle and apply the vaccine in this way. It appears that the skills are being readily transferred to producers that catching birds is not a problem, and that subsequent vaccinations will be handled by producers themselves, with only limited input from veterinary staff.

It is clear that the potential risk of Newcastle disease to the poultry industry in Zimbabwe will ensure the issue remains a high priority. Equally clear is the need to sustain producer commitment and cooperation if the disease is to be controlled in the extensive poultry flock. Newcastle disease represents only one of many constraints to the backyard sector in Zimbabwe. Other health, management and husbandry limitations have been largely ignored by the service sector, and community support for Newcastle disease control will depend on the commitment of the service sector to meet this much broader range of needs.

Neoplastic Diseases in Poultry

Virus-induced Neoplastic Diseases Marek's Disease

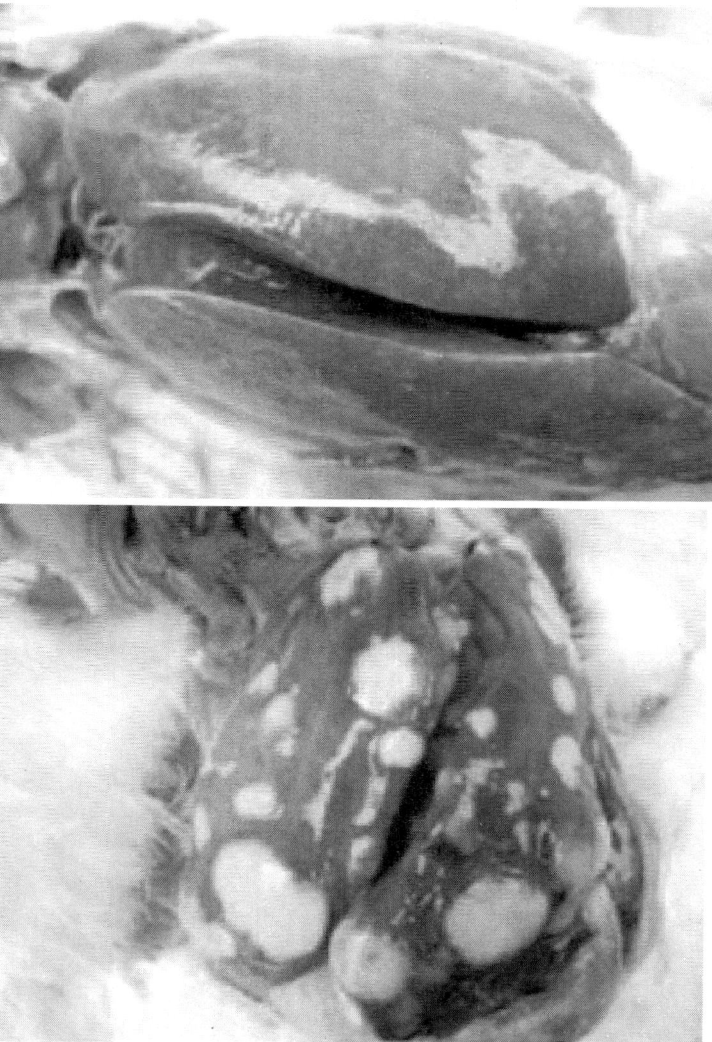

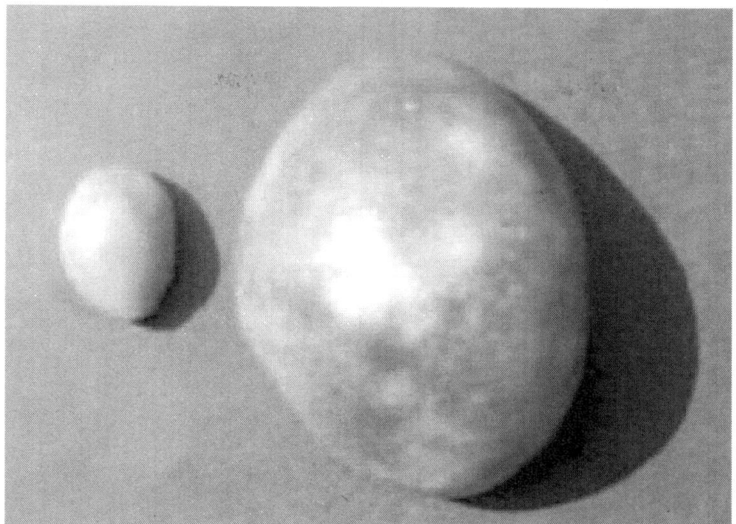

279.280.281. Acute (visceral) form. It is characterised by diffuse or nodular lymphomatous lesions in various viscera (liver, spleen, heart, kidneys, lungs, gonads, proventriculus, pancreas etc.), the skeletal muscles and the skin. MD affects mainly hens, and is rarely observed in turkeys. It is most commonly encountered in birds at the age of 89 weeks and in layer hens. The cases at the age of 16-20 and 24-30 weeks are predominant. MD is prevalent all around the world and in fact, all flocks are exposed to the effect of the aetiological factor.

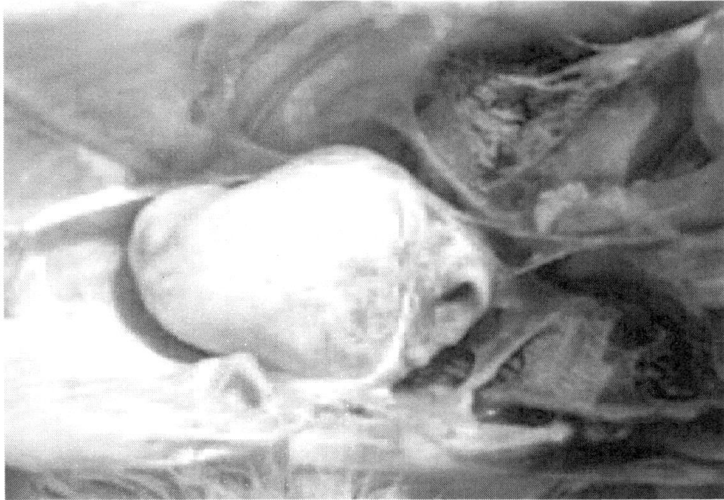

282. Diffuse lymphomatous growths in the heart, resulting in its transformation into an amorphous tumours mass.

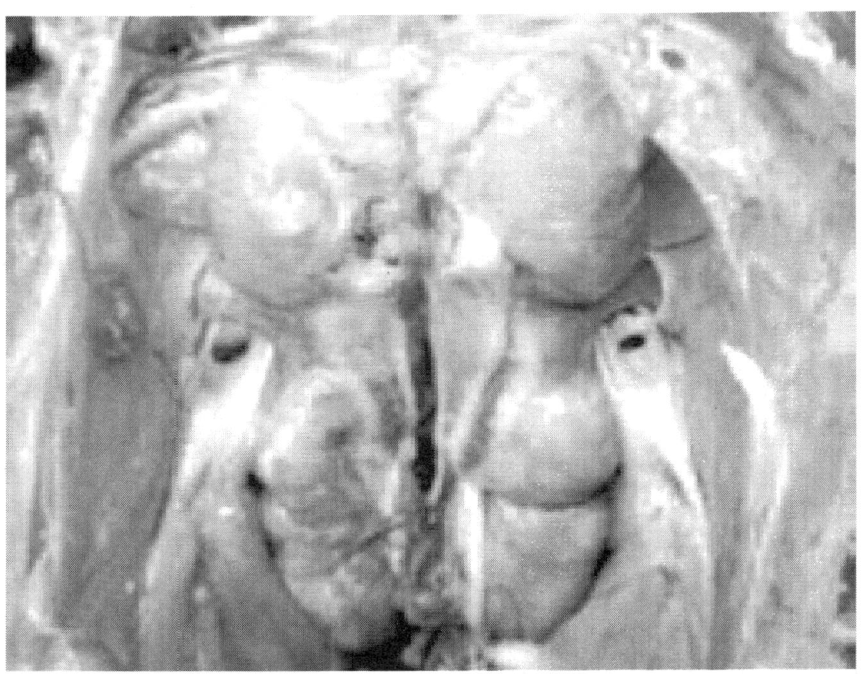

283. Bilateral enlargement of kidneys because of a diffuse lymphoid cell proliferation.

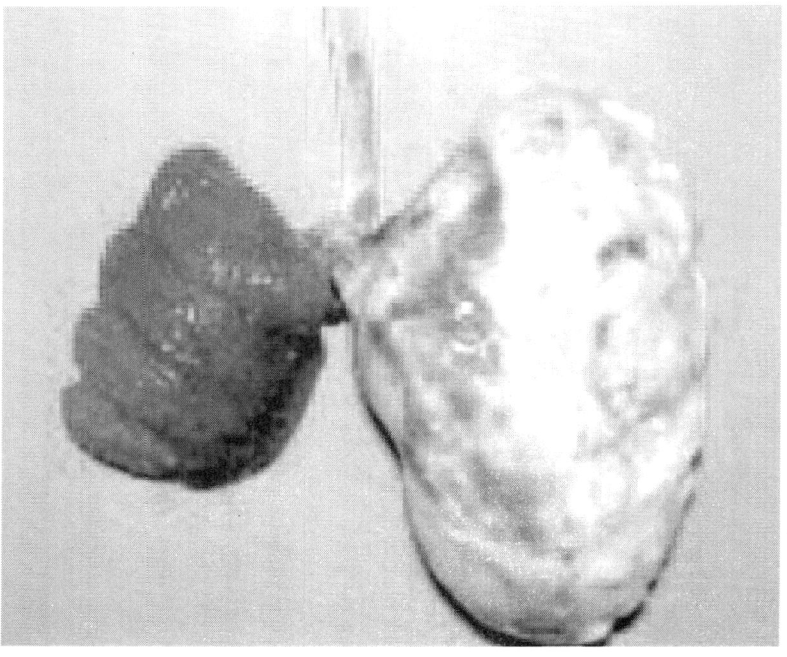

284. Neoplastically modified right lung in MD.

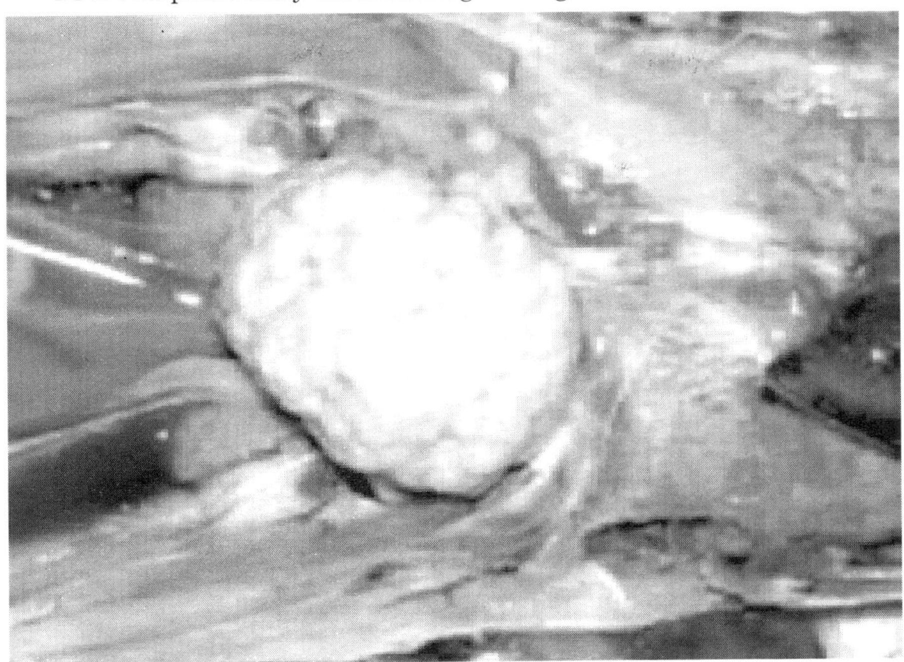

285. Typical cauliflower-like appearance of the ovary, distinctive for MD.

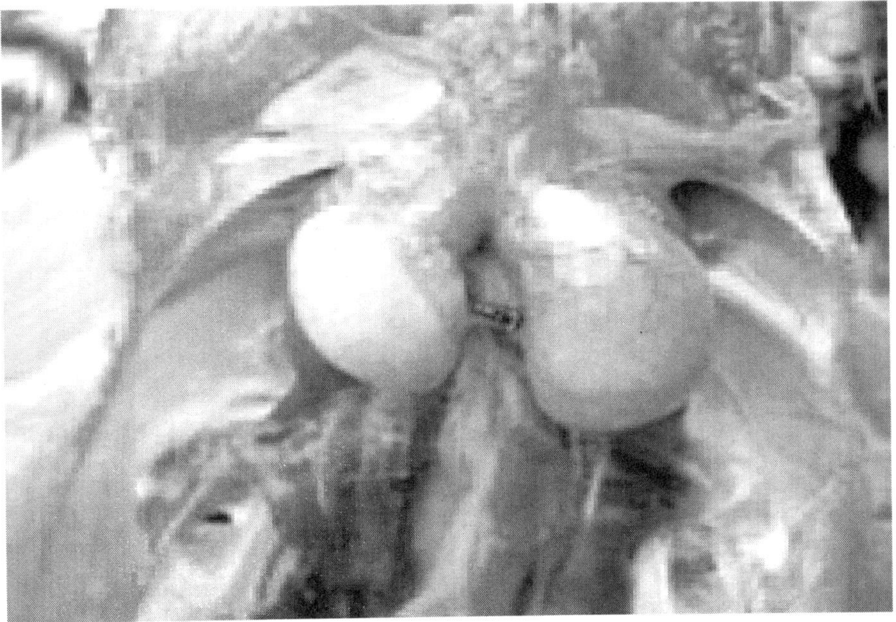

286. Marked asymmetry of testes on a cock following unilateral lymphoid cell proliferation.

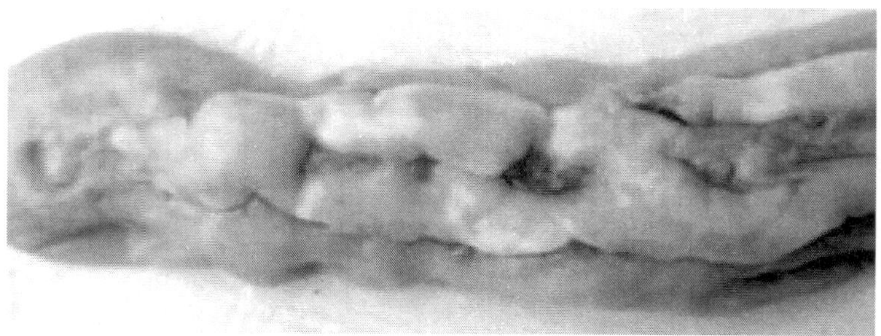

287. Diffuse neoplastic growths affecting the pancreas.

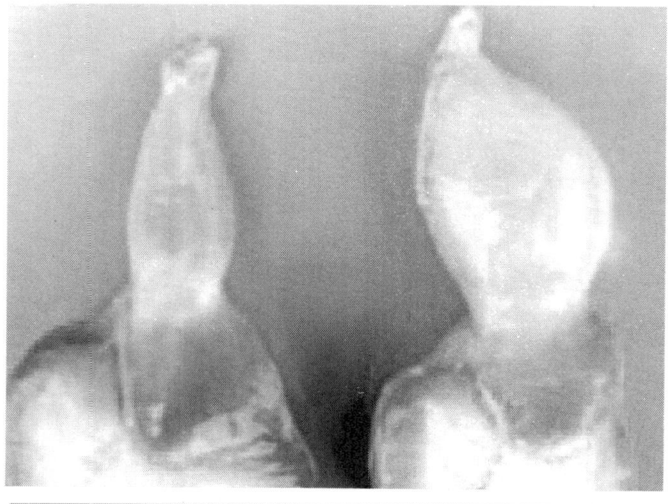

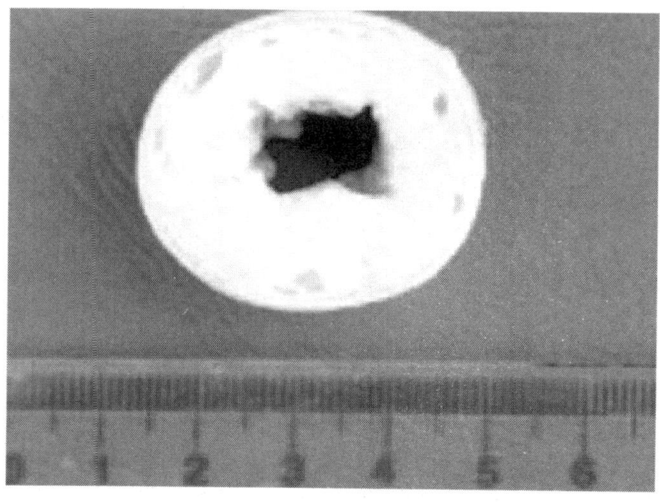

288.289. The manifold enlarged proventriculus with the shape of a round bottom flask (288) result of diffuse neoplastic growth and the severely narrowed lumen (289) are a typical finding in MD. The causative agent of MD is a type B cell associated herpesvirus (MDV). There are three MDV serotypes. The isolates of serotype 1 are widely distributed among hens and vary from highly virulent (w+) oncogenic to almost avirulent strains. The serotype 2 is common for hens and is not oncogenic. The isolates of serotype 3, known also as turkey herpesviruses (HVT) are naturally occurring in turkeys and are non-oncogenic. The three serotypes possess a significant cross reactivity.

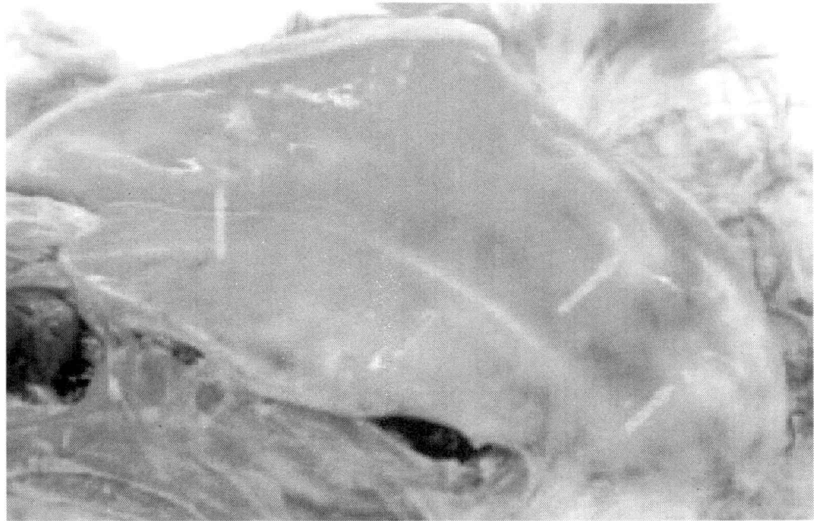

290. Multicentric MD tumours (arrows) prominating or seen through the superficial and deep pectoral muscles.

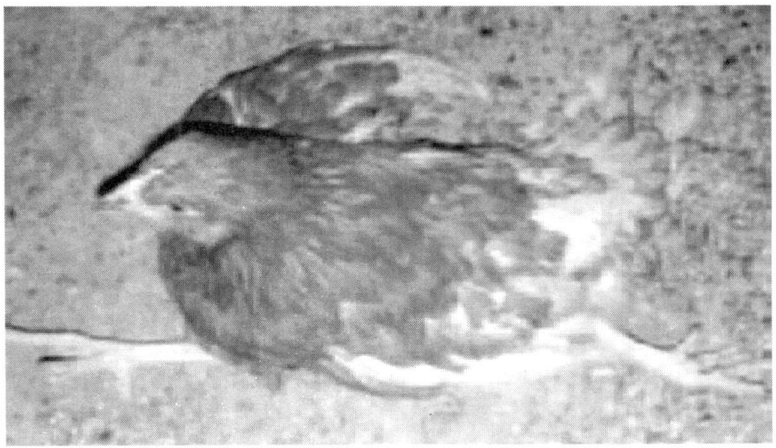

291.292. Chronic (classical) form. It is encountered as neural type (fowl paralysis) or ocular type (ocular lymphomatosis). Clinically, the neural for is manifested with paralysis of limbs.

293. The ocular form is characterised with iris depigmentation, deformation of the pupil, sometimes opacity of the cornea and blindness.

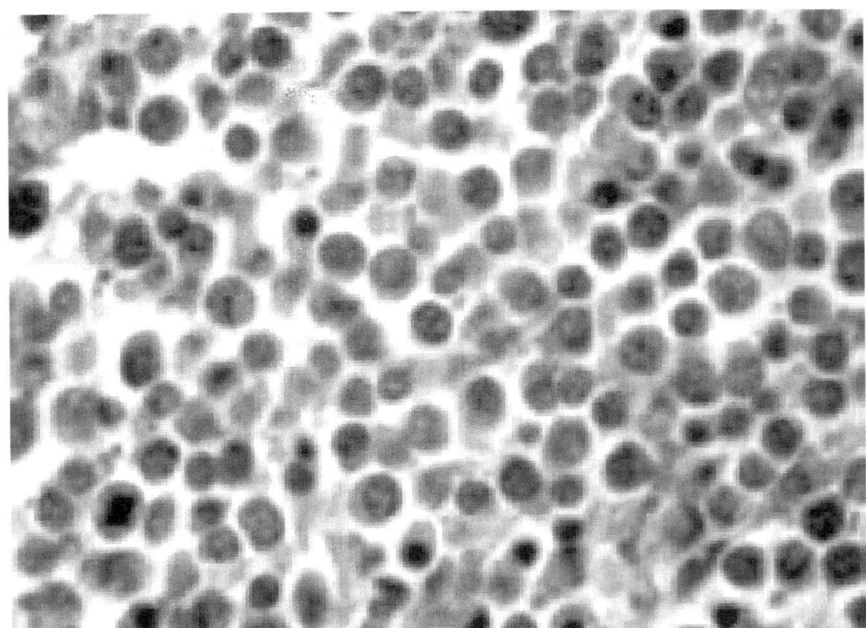

294. Histologically, pleiomorphic lymphoid cell proliferations in affected viscera, nerves or eyes are observed.

The Marek's disease is probable provided that at least one of the next conditions is present: peripheral nerves augmentation, depigmentation of the iris or irregularly-shaped pupil; lymphoid tumours in various organs in birds younger than 16 weeks; presence of visceral tumours in birds at the age of 16 weeks and older; simultaneous lack of alterations in the bursa of Fabricius.

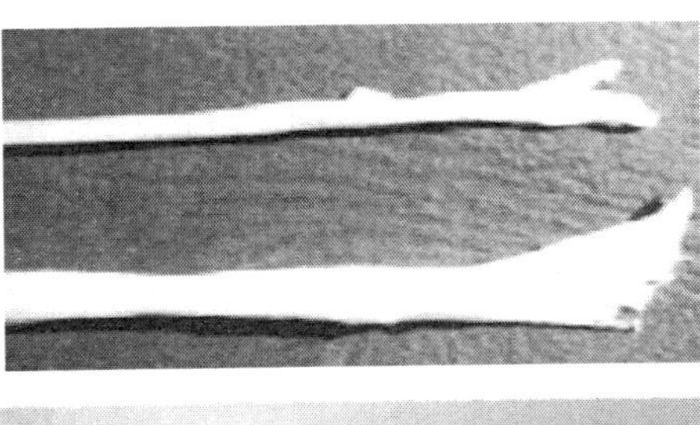

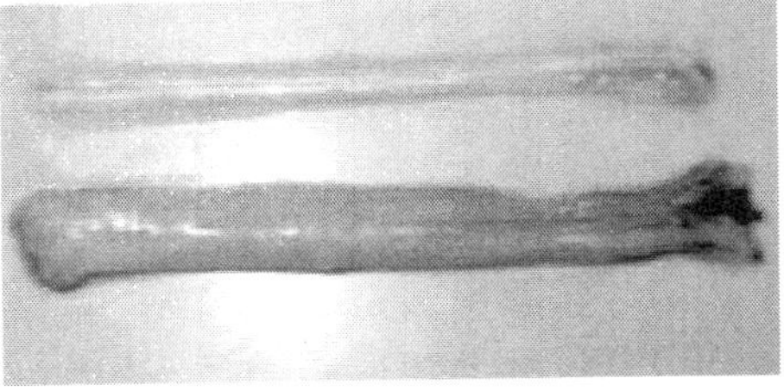

295.296. Pathoanatomically, unilateral or bilateral thickening of affected nerves, mainly diffuse and at a various extent, is observed.

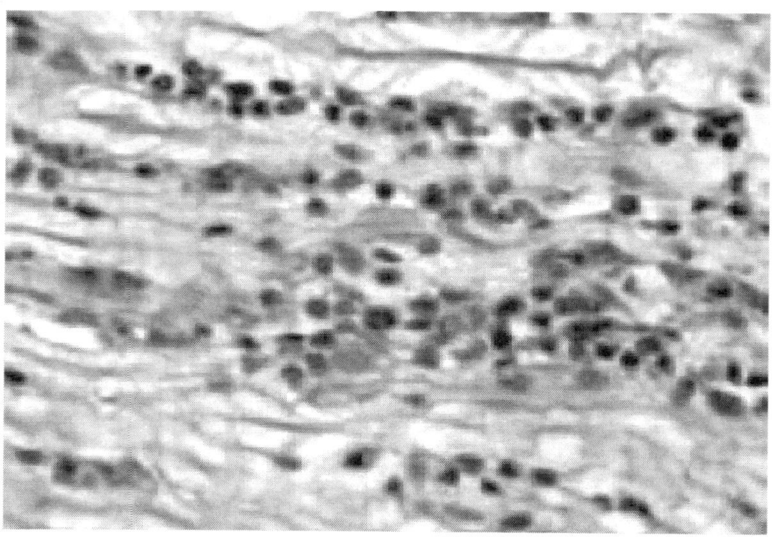

297. Microscopical view of lesions in a peripheral nerve consequent to MD.

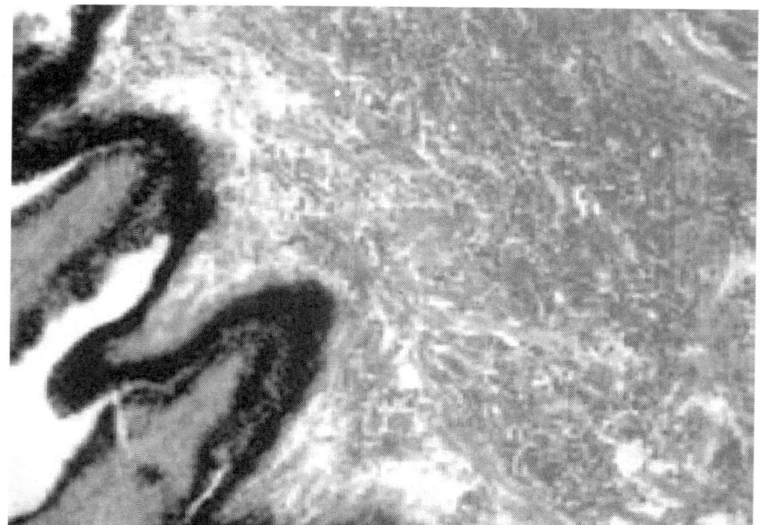

298. Lymphoid cell proliferations in the iris and the ciliary muscles in the ocular form of MD. Three classes of viruses are able to protect fowl from MD: attenuated serotype 1 of MDV-cell associated vaccines; HVT could be used for preparation of cell-free lyophilised vaccines;

naturally apathogenic isolates of serotype 2 cell associated vaccines. The vaccines against MD provide over 90% protection. HVT gives excellent results but in case of failure, a bivalent vaccine could be used.

299. Transient paralysis. They are observed in chickens and hens, especially non-vaccinated against MD. Most cases present the classical form manifested by flaccid paralysis of the neck and legs for 143 days followed by complete recovery. The syndrome has to be differentiated from the neural form of MD on the basis of its transient nature and the flaccid, but not spastic paralysis.

lymphoid Leukosis

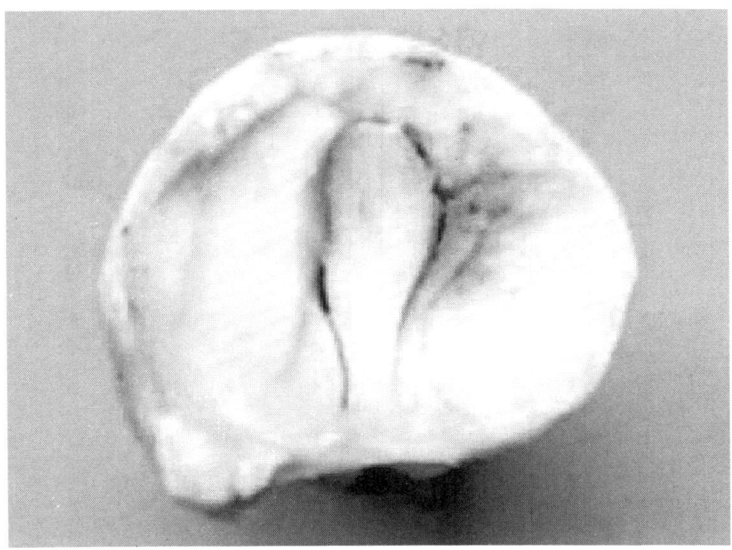

300.. It is characterised by a gradual beginning, persistent low mortality in the flock and diffuse or focal neoplastic growths of lymphoblasts in viscera. The neoplastic changes begin always from the bursa of Fabricius, where various-sized lymphomas are detected (transverse section through neo-plastically grown bursa -fixed preparation).

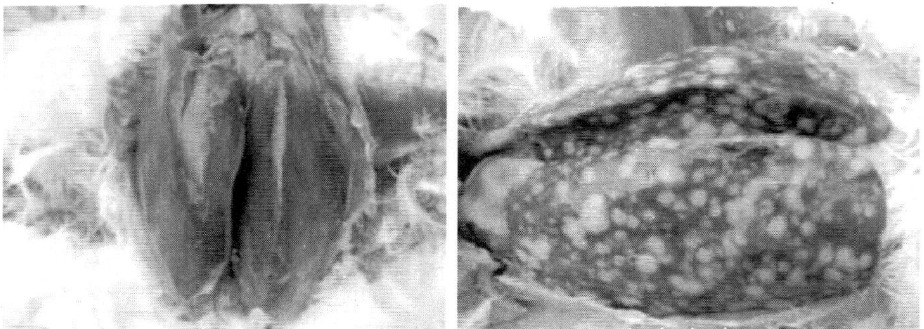

301.302. Clinically, pale comb and wattles, sometimes swelling of the abdomen because of the highly enlarged liver are observed. Diffuse or nodular neoplastic growths could be detected in many organs, but they are more common in the liver, the spleen, the kidneys, the heart and the ovary.

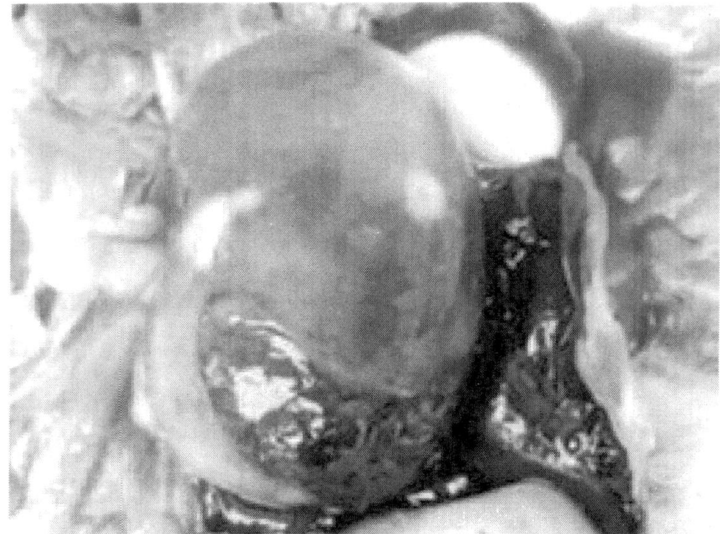

303. Spontaneous rupture of the neoplastically grown spleen, leading to extensive loss of blood. LL is widely distributed worldwide in countries with developed industrial poultry breeding. It is usually observed in birds at the age of 16 weeks and older.

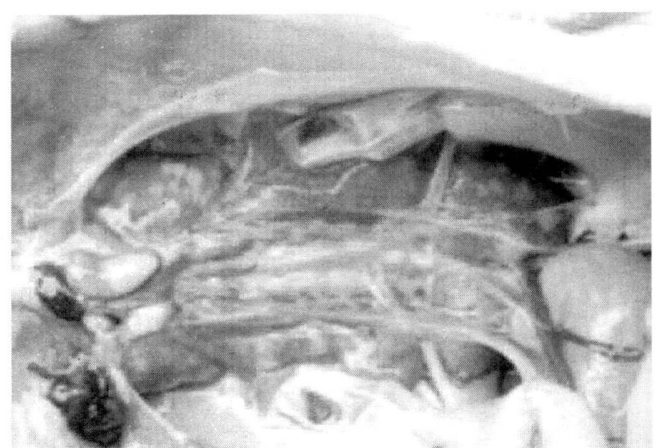

304. Focal neoplastic lesions in kidneys. LL is caused by viruses of the L/S group classified in 10 subgroups: A, B, C, D, E, F, G, H, I and J. The viruses from subgroup A are most prevalent and most frequently associated with LL. Hens, rarely turkeys, pheasants and quails are susceptible.

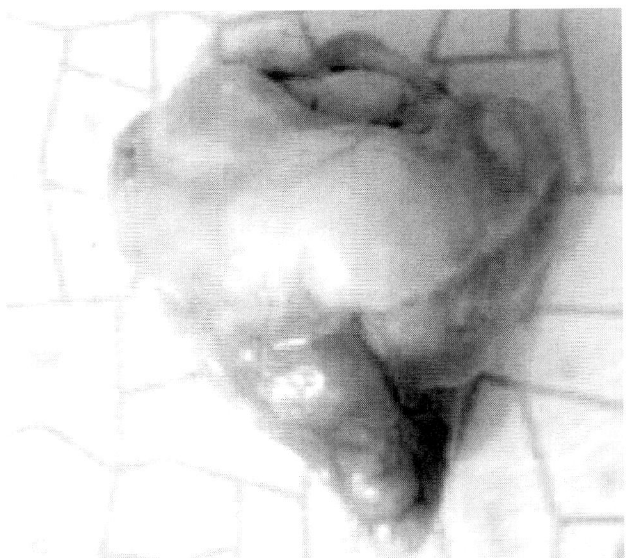

305. Diffuse and focal tumour lesions in the heart. The replication of the virus occurs in albumin secreting glands of the oviduct. The transmission of the infection is performed vertically by egg albumin from one generation to another. The role of cocks is not important for the congenital infection of the progeny. They are only virus carriers and source of venereal infection for other birds.

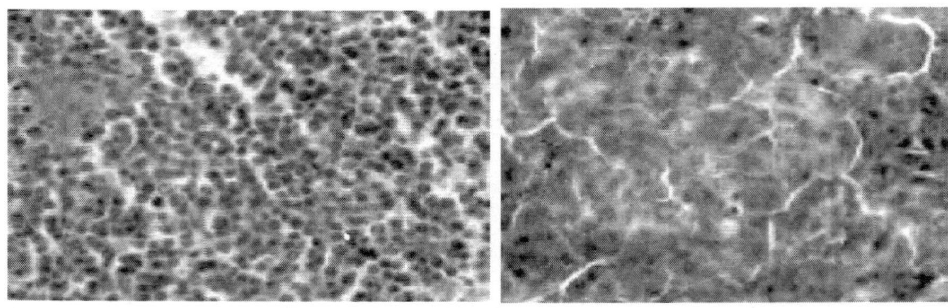

306.307. Histologically, growth of single type lymphoblast cells with marked pyroninophilia is observed.

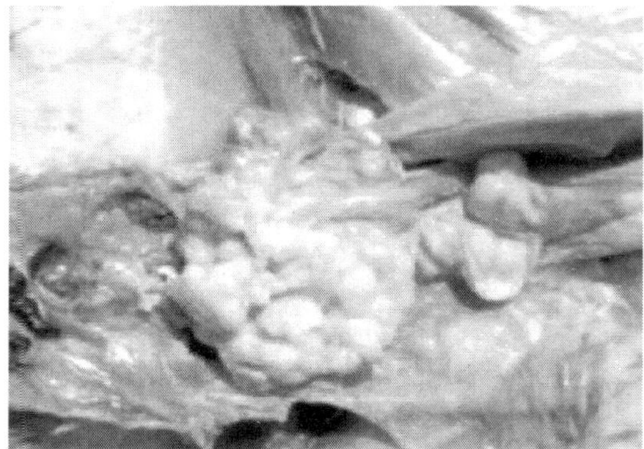

308. Neoplastically transformed ovary in LL. In some instances, the horizontal infection is also possible but only in chickens in the first few days after hatching, usually via vaccines contaminated with ALSV. The lethal issues are observed for 56 months after the LL outbreak and amount to 5 -15%.

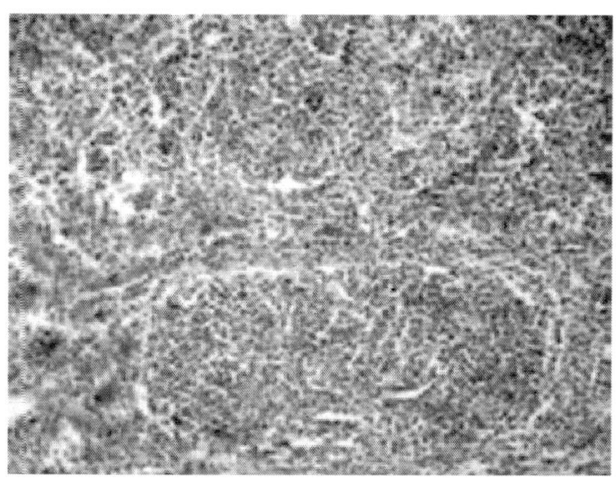

309. In the bursa of Fabricius, a characteristic intrafollicular hyperplasia is observed.

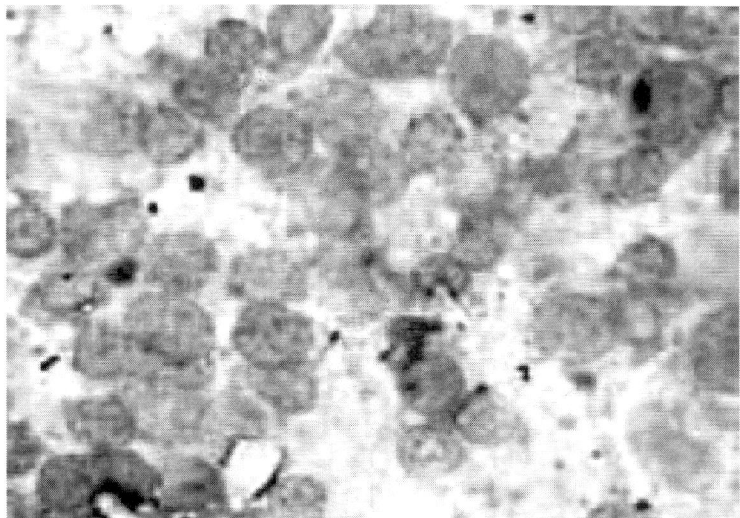

310. The picture of an imprint preparation from neoplastic lesions shows a layer of single-type lymphoblast cells. LL and MD are hard to be distinguished: in both, lymphoid tumours are present in the same visceral organs, the appearance at the same age is possible, and the visceral lesions could not be differentiated macroscopically, except in a careful microscopic examination by an experienced pathologist.

From the Point of View of Differential Diagnosis, the Following Features Deserve a Special Emphasis

Lymphoid Leukosis

- Usually, it is not seen in birds younger than 14 weeks.
- The lethal issues occur mostly at the age between 24 and 40 weeks.
- Distinct nodular tumours.
- Tumours in the bursa of Fabricius.

Marek's Disease

- Could be observed after the age of 4 weeks too.
- The peak mortality is seen between the 10th and the 20th week, sometimes continues after the 20th week.
- Paralysis.
- "Grey eye".
- In some birds, the bursa of Fabricius is atrophied, in others: neoplas.

Myelocytomatosis

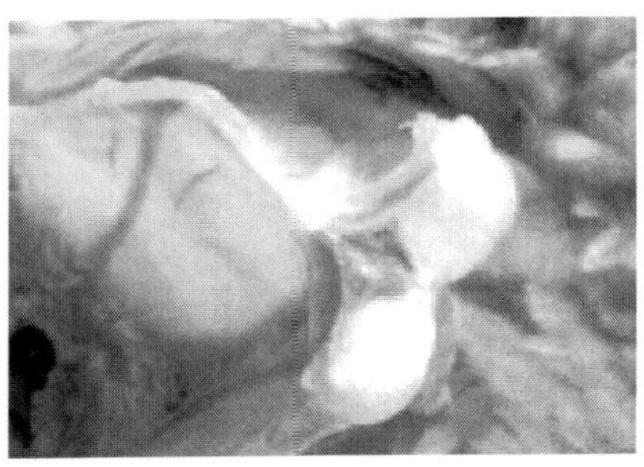

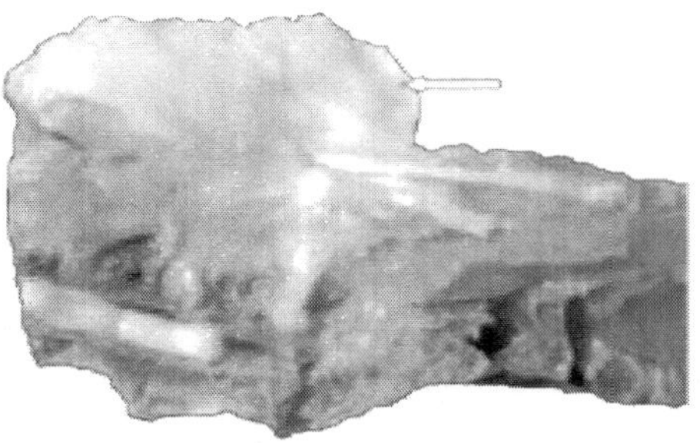

311.312.MC is caused by viral staings of ALSVs from sungroups A,B and J (Mc29, MC31, Cm11, OK10, HRPS 103, and ADOL HC1). It is encountered relatively infrequently. Its occurrence is sporadic or enzootic. Suceptible birds and hens, pheasents, guinea hens and quails. In most cases, the liver is enlarged, thick and mottled with dark red sports or fat like nodules.

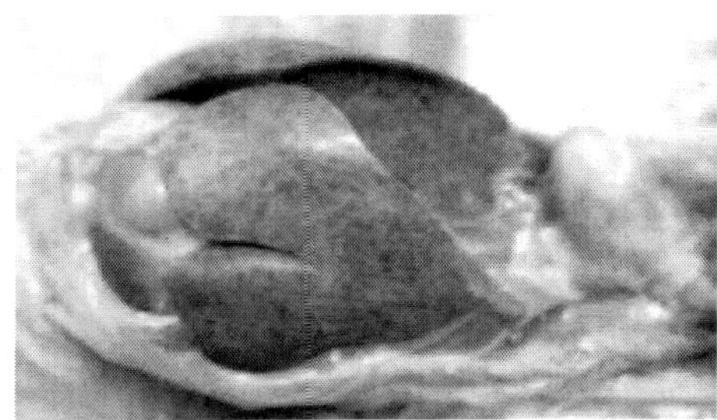

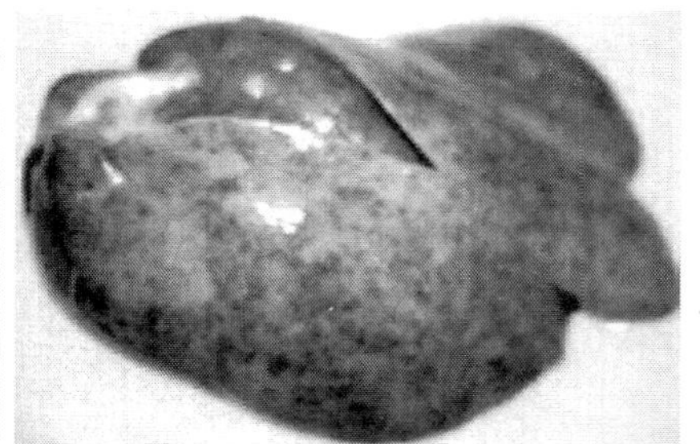

313.314.. MC is caused by viral strains of ALSVs from subgroups A, B and J (MC29, MC31, CMII, OK10, HRPS 103, and ADOL HC1). It is encountered relatively infrequently. Its occurrence is sporadic or enzootic. Susceptible birds are hens, pheasants, guinea hens and quails. In most cases, the liver is enlarged, thick and mottled with dark red spots or fat-like nodules.

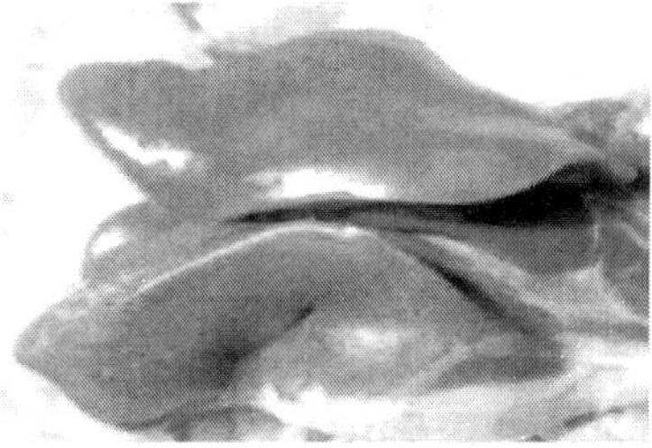

315.Sclerotic changes in the liver are possible because of regression of neoplastic lesions.

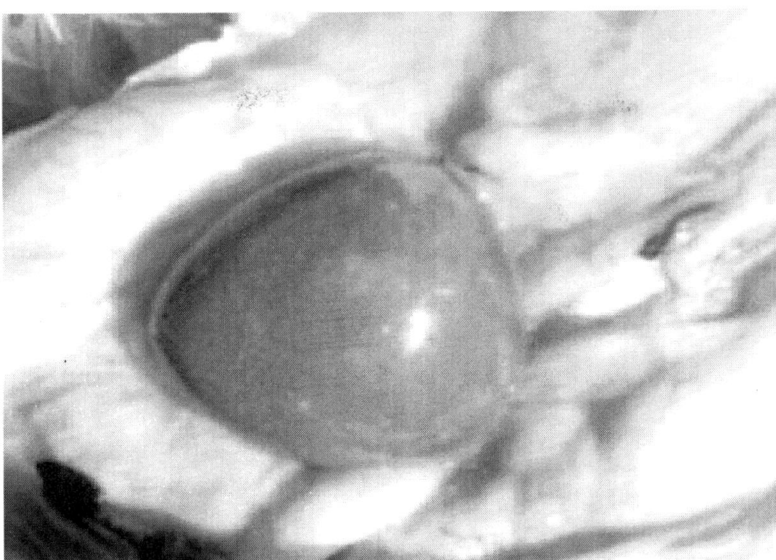

316.The spleen is usually enlarged, but sometimes, could be atrophied.

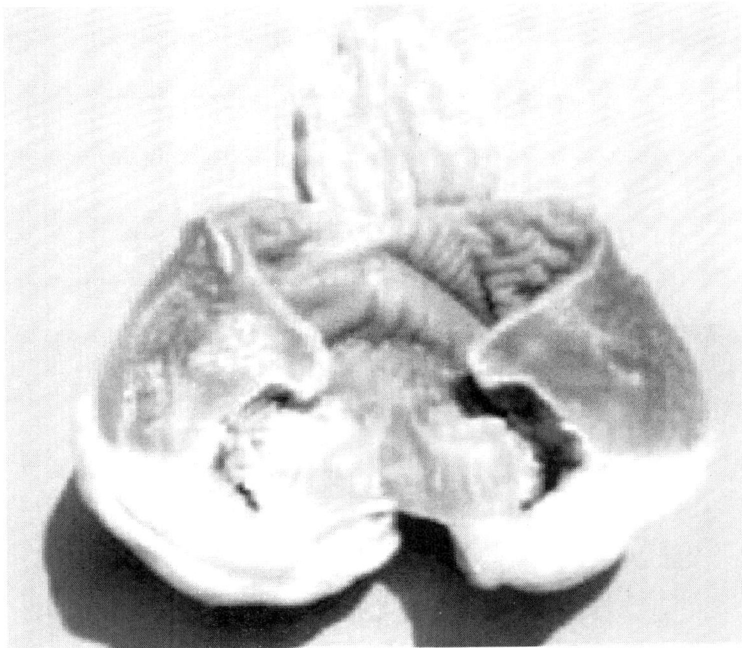

317.. A characteristic feature of MC is its simultaneous course with tumours from a different type: mesenchymal, epithelial or mixed. The picture shows a fibrosarcoma to the gizzard associated with MC.

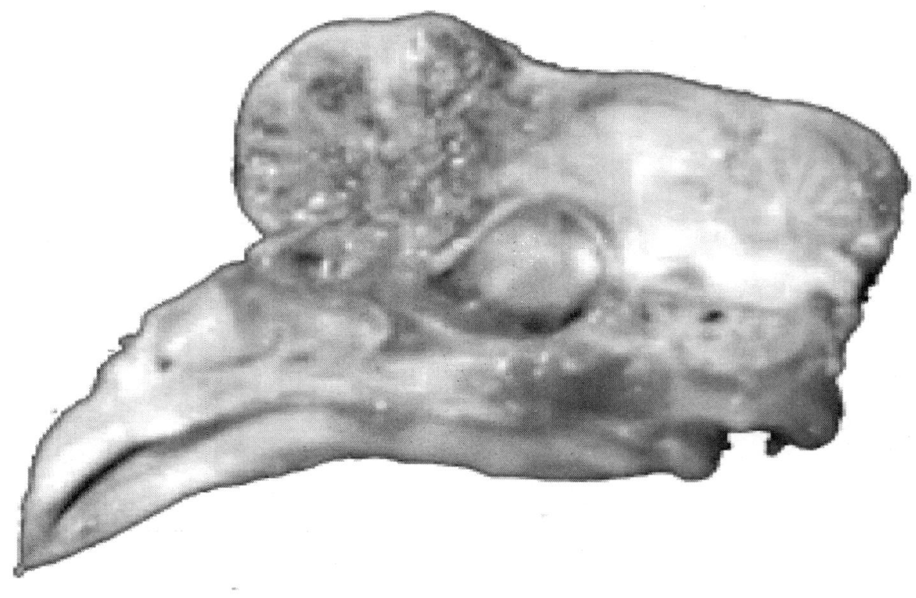

318.Mixed mesenchymal tumour (osteochondrosarcoma) to the frontal skull bones: a sagittal cross section

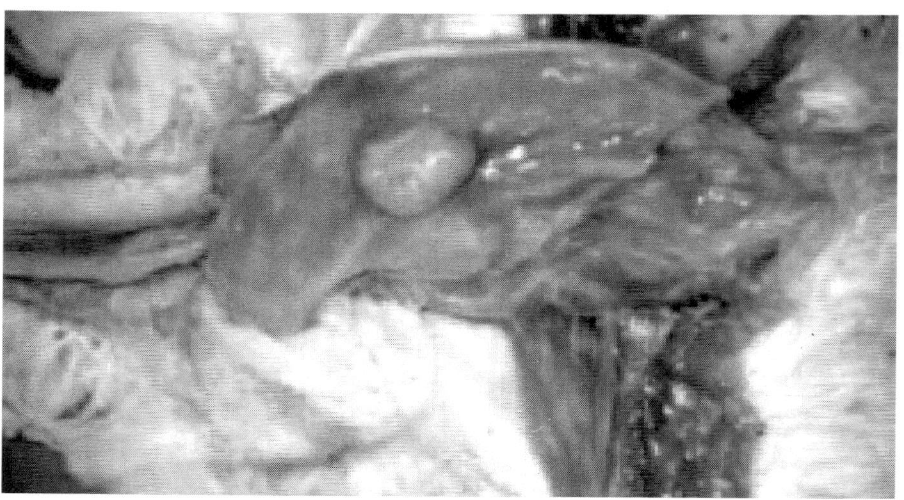

319.Multiple rabdomyosarcoma in pectoral, thigh, abdominal and tracheal muscles.

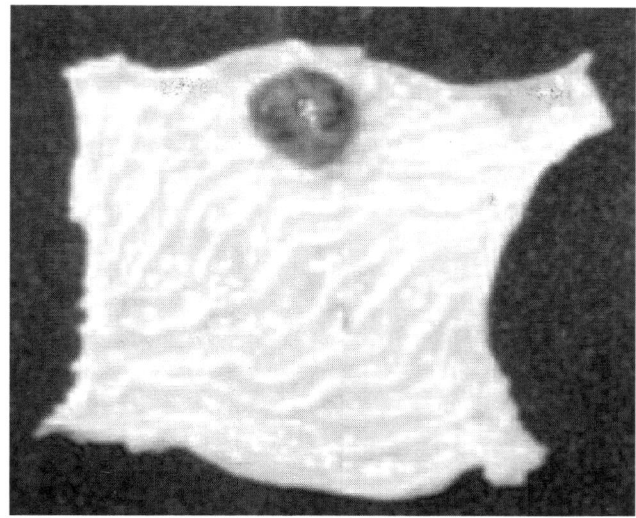

320.Leiomyosarcoma of the mucous coat on the oviduct.

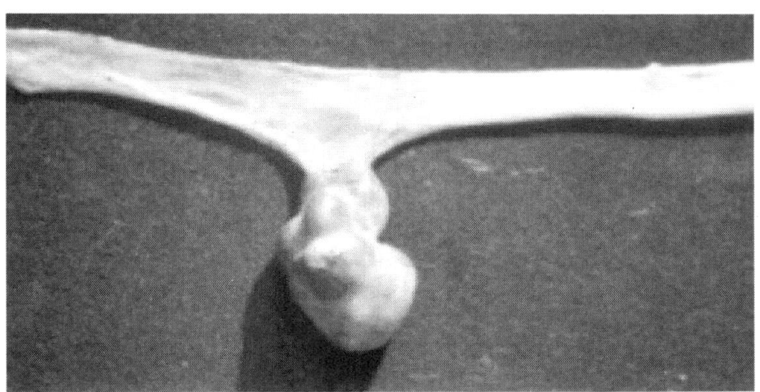

321.Pendulation harmangio-sarcoma of the ileal serosa.

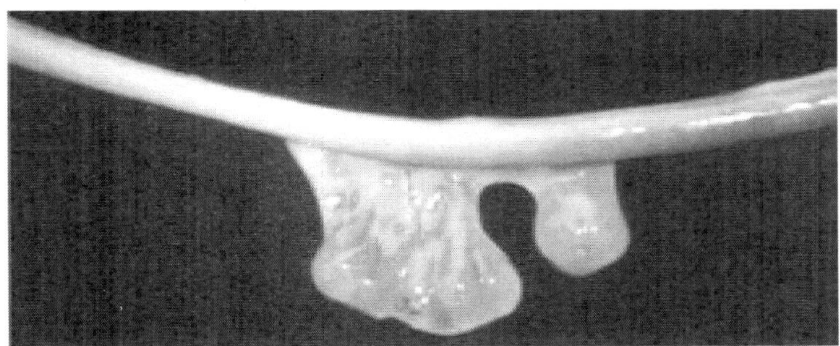

322.Pendulating multiple myxoma of the small intestine's serous coat.

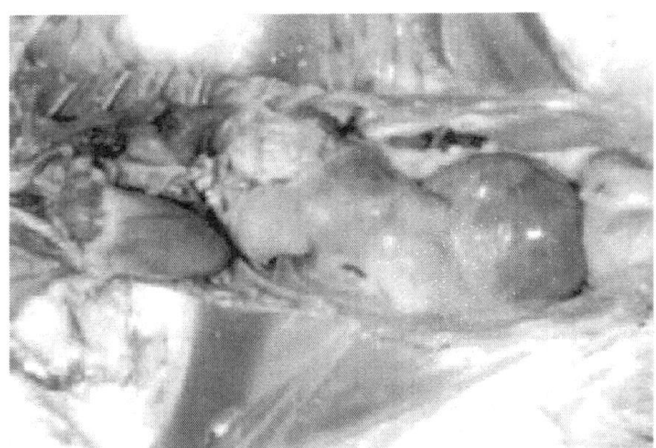

323.. MC-associated cystadenocarcinoma of the kidney in a hen.

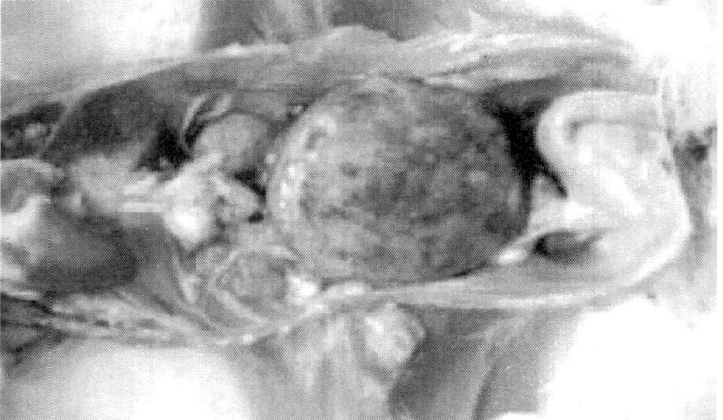

324.Nephroblastoma of the left kidney, occupying a significant part of the abdominal cavity.

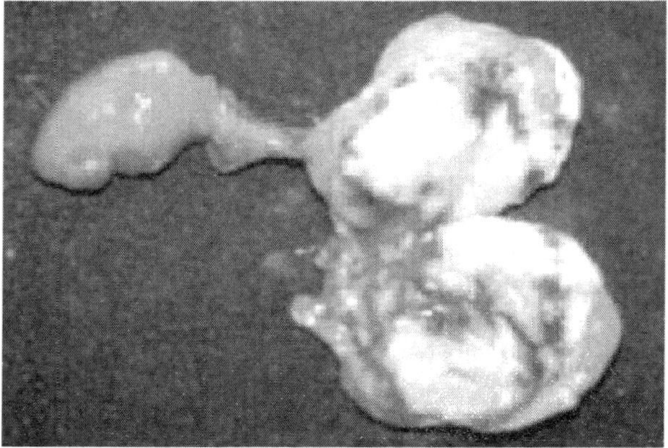

325.Nephroblastoma -the surface of a cross section. The tumour is a pendulating mass attached to the kidney by a fibrous vascularized stem that has undergone a partial necrosis and haemorrhages

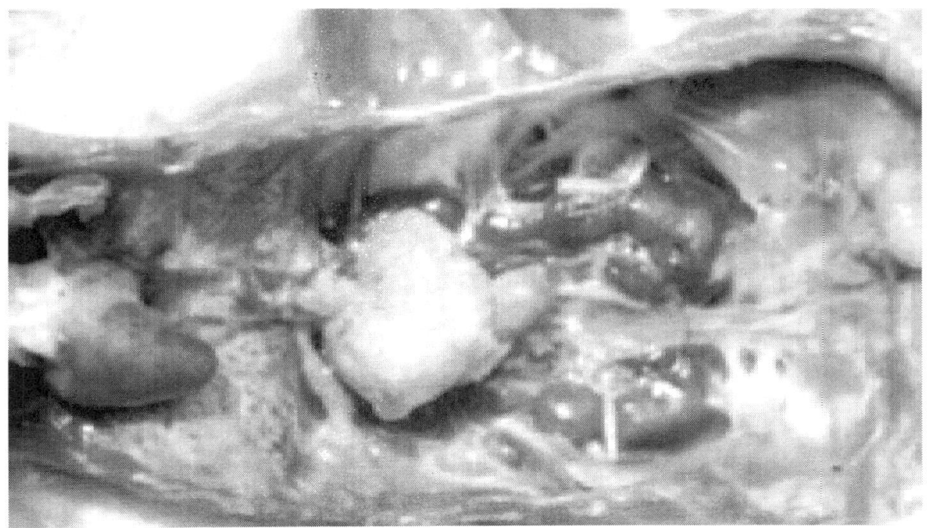

326.Granulosa cell tumour of the ovary. The tumour appears as a single, compact, dorsoventrally flattened growth.

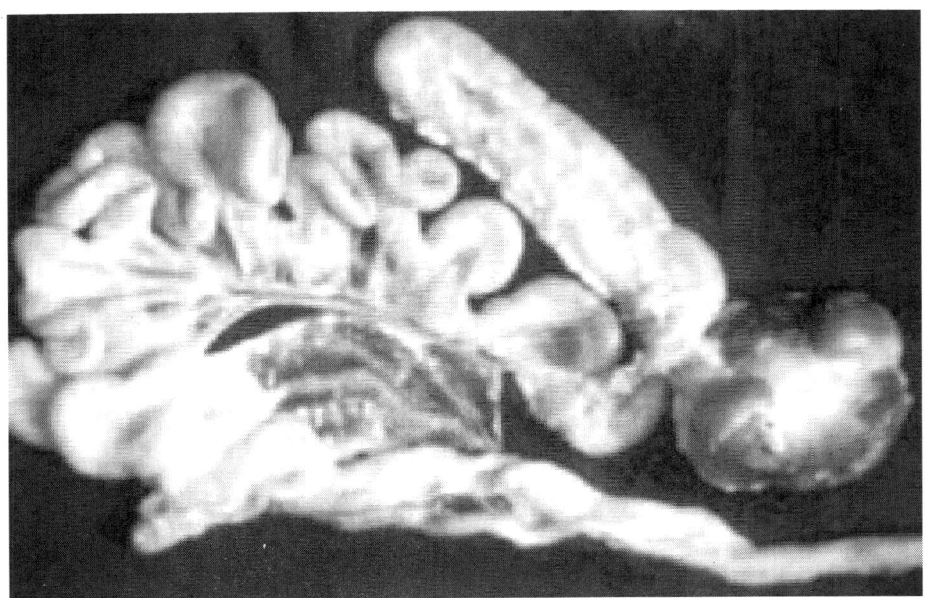

327.MC-associated multiple carcino¬sarcoma of the mesentery and alimentary tract's serous coat (disseminated milliary nodules).

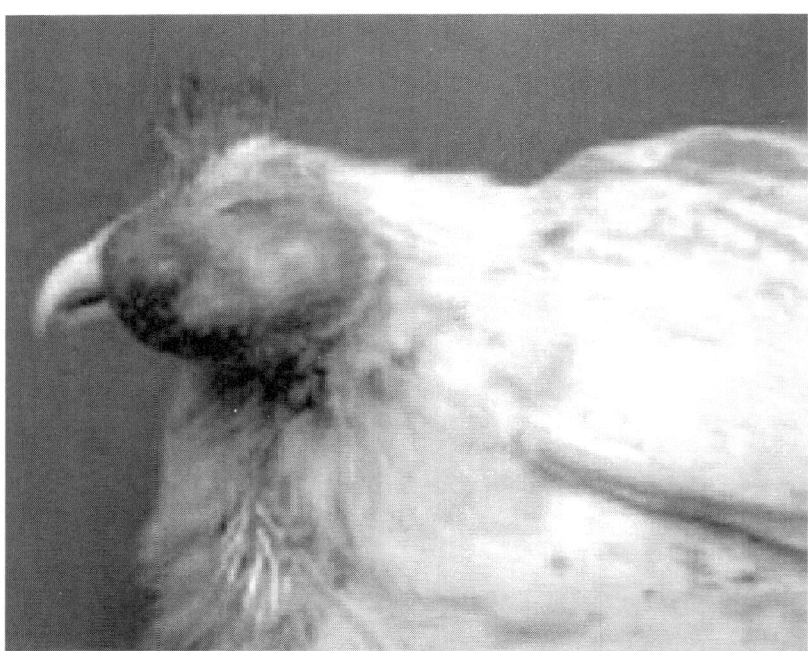

328.MC-associated carcinosarcomas in the region of the right infraorbital sinus.

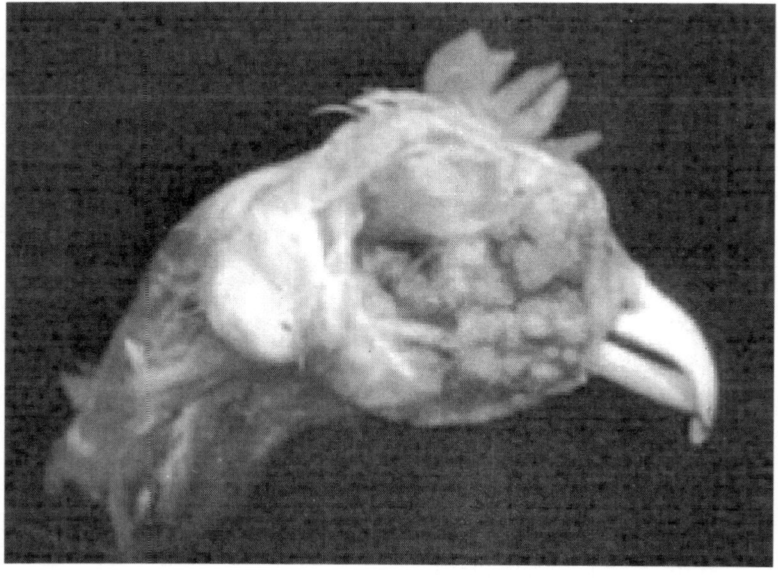

329.Gross appearance of the tumour from Fig. 328 after removal of the covering skin.

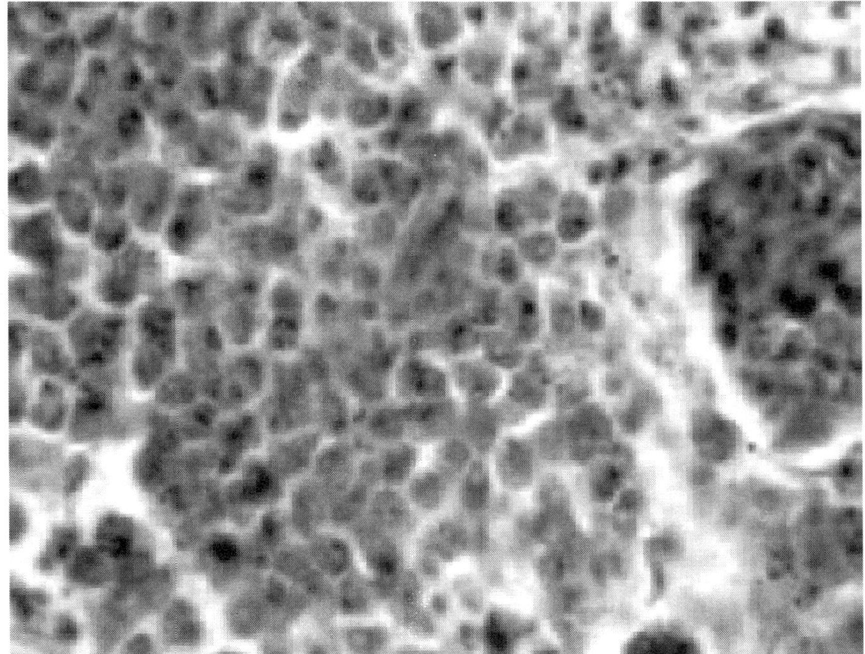

330.Histologically, myelocytomatomas are easily distinguished. Most commonly, they have perivascular localization. Growth of myelocytes with well-formed granules in a liver cross-section.

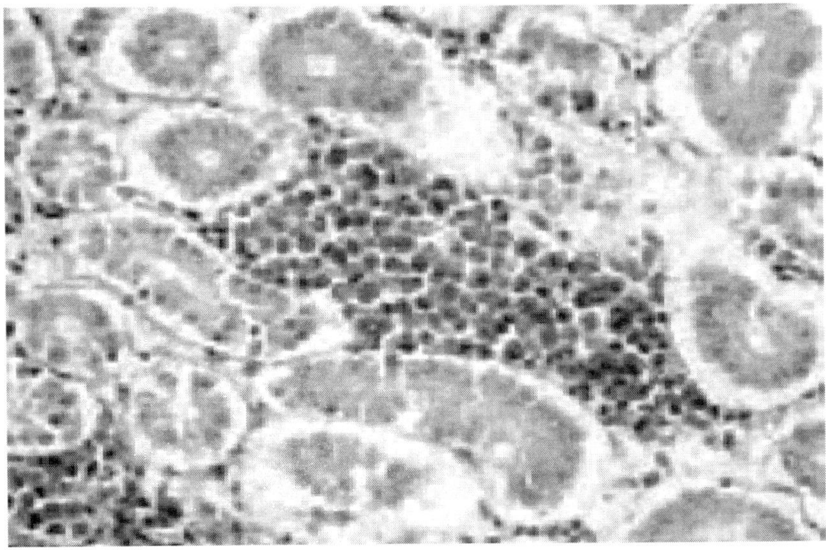

331.Kidney.Focal intertubular myelocytic pro¬liferations.

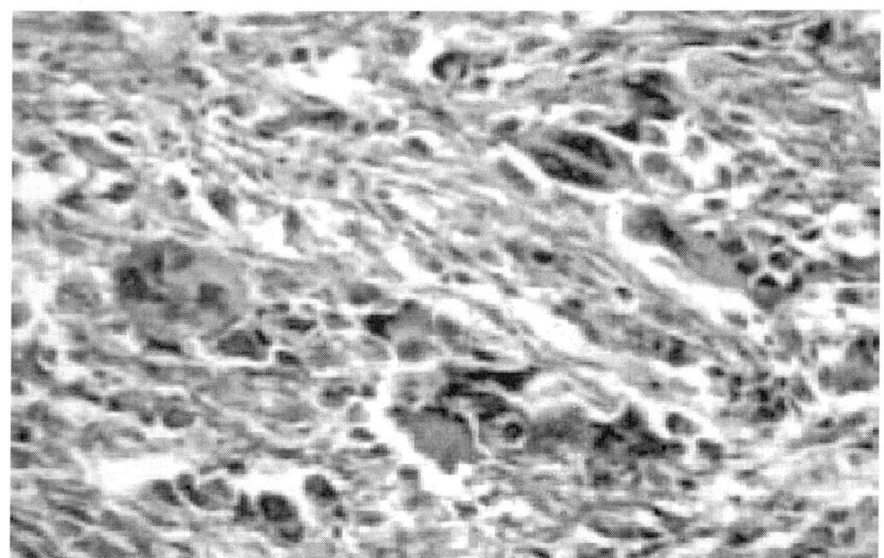

332.. MC-associated neoplasms of epithelial, mesenchymal or mixed type demonstrate the respective type of histological structure. Leiomyosarcoma a histological view. Polygonal giant cells with hyperchromatic nuclei.

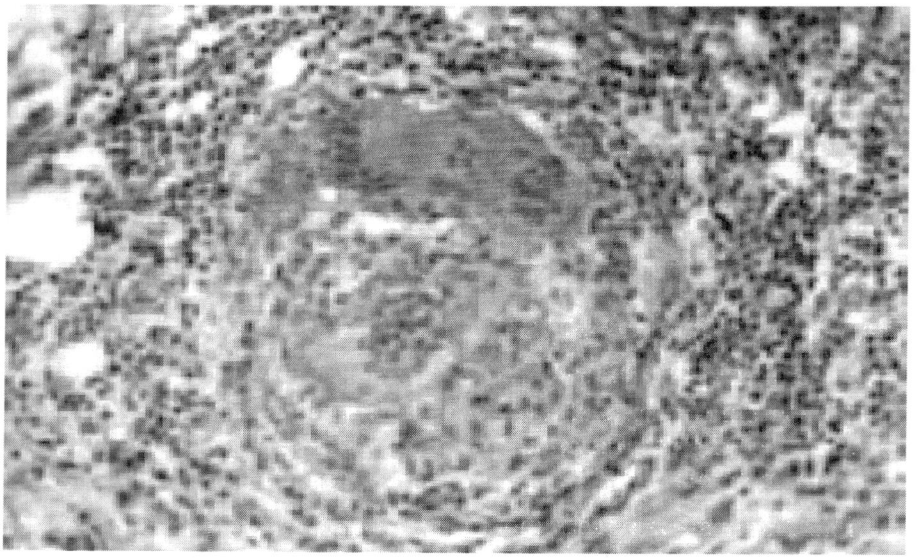

333.Leiomyosarcoma - small intestime. Prolongations of polynuclear symplastic elements.

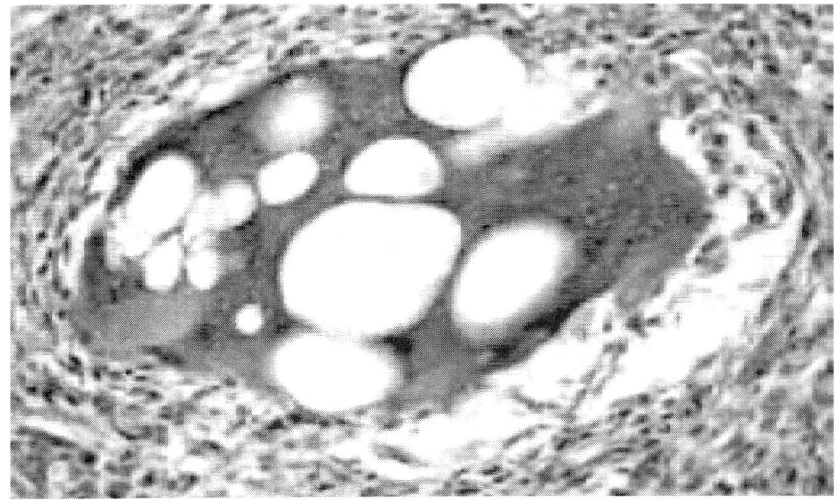

334.Leiomyosarcoma - small intestine. Extraordinary („monstrous") multinuclear giant cell with intracytoplasmic vacuoles.

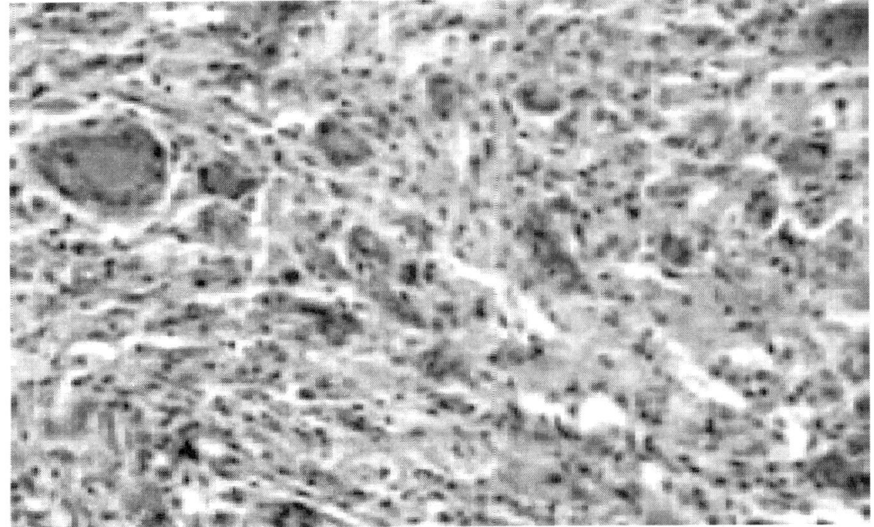

335.Carcinosarcoma of the pancreas. Tubulous glandular epithelial formations of the carcinoma component among the liposarcoma part of the parenchyma.

The diagnosis is based upon the entity of data about the history, the gross appearance and location of the tumours and the specific histo¬logical lesions. From a differential diagnostic point of view, myelo¬blastosis and erythroblastosis should be considered.

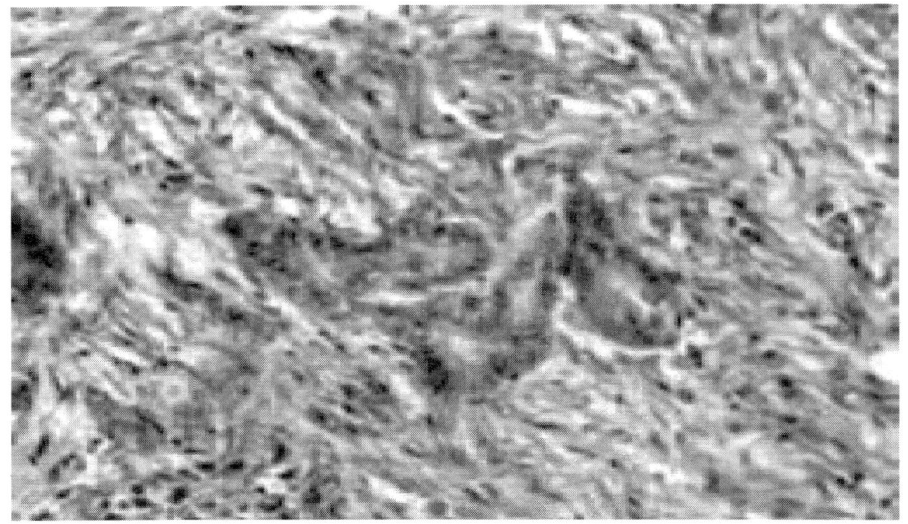

336.. Rabdomyosarcoma. An area with multiple hyperchromatic giant cells.

Erythroblastosis

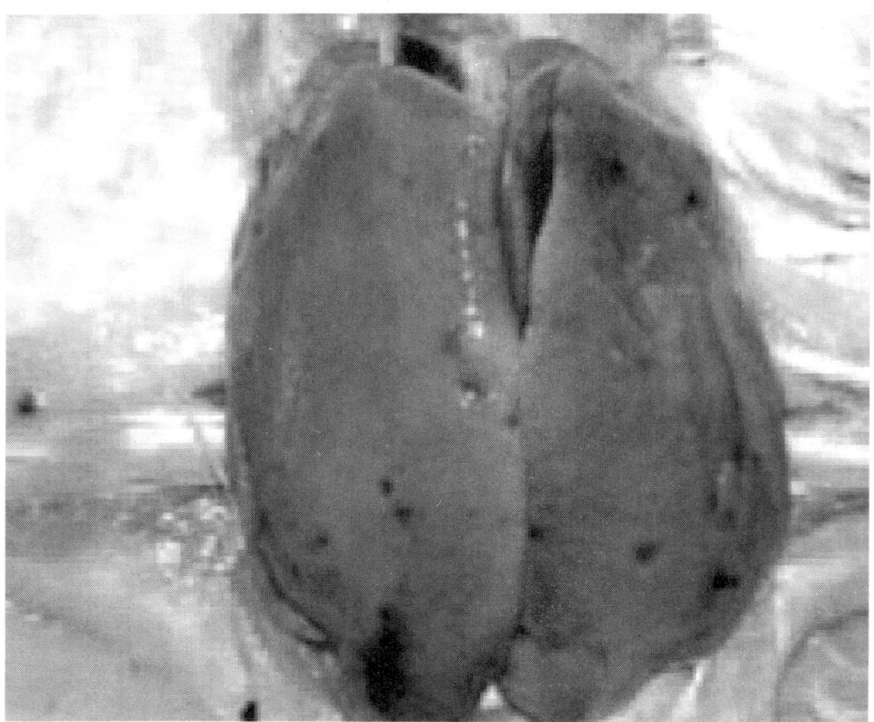

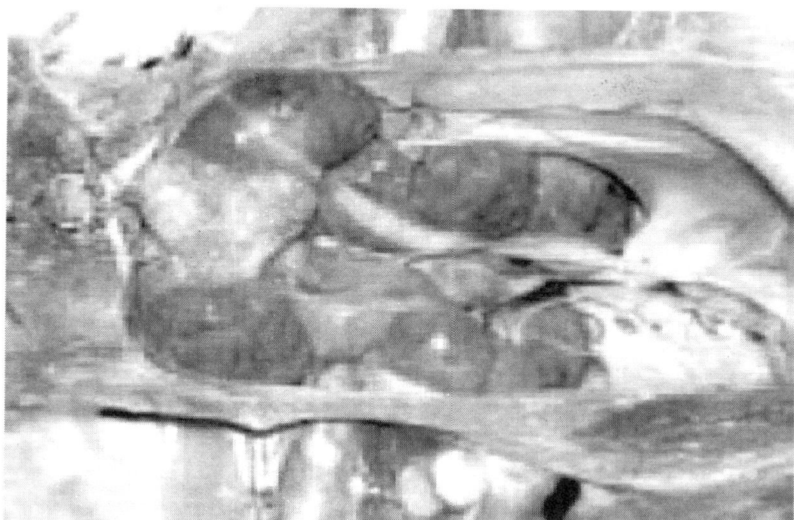

337.338.Erythroblastosis (ER) is characterized by intravascular proliferations of immature precursors of erythrocytes. ER has a leukaemic character and is manifested with signs of severe anaemia. The liver and the kidneys are moderately enlarged with a characteristic dark red to mahogany colour, sometimes with haemorrhages.

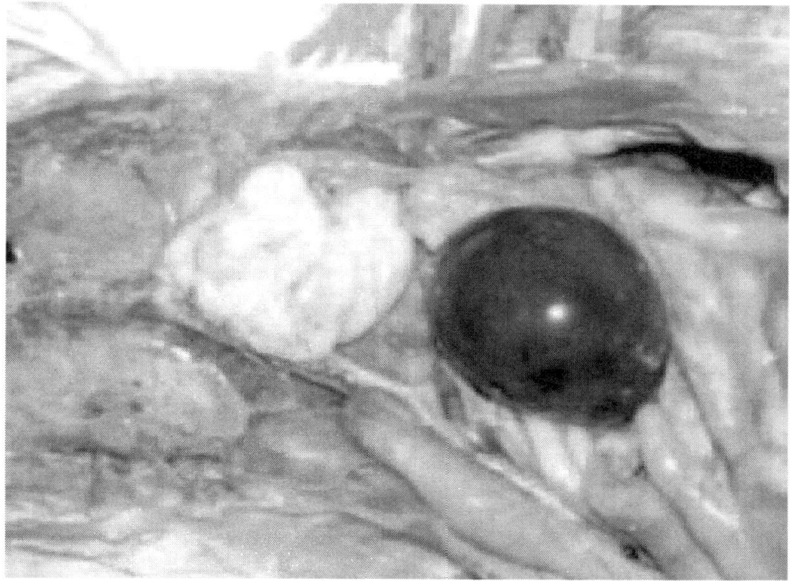

339.ER is caused by the avian eryhtroblastosis virus (AEV); the most frequently encountered strains are E-26, ES4, R etc. The spleen is unusually enlarged or atrophied in cases of severe anaemia.

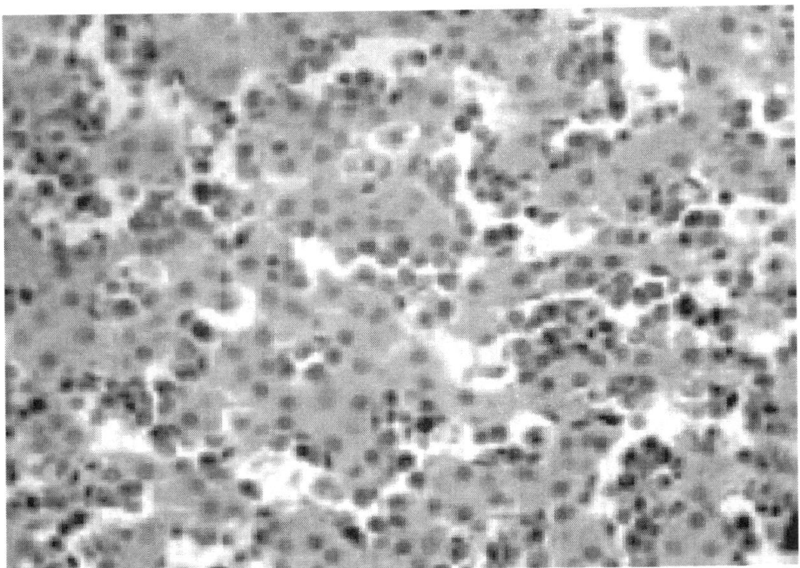

340.Histologically, accumulation of erythroblasts in blood sinusoids and capillaries is seen. The diagnosis is based on visceral histological lesions, typical for ER and peripheral blood haematological and morpho¬logical analysis.

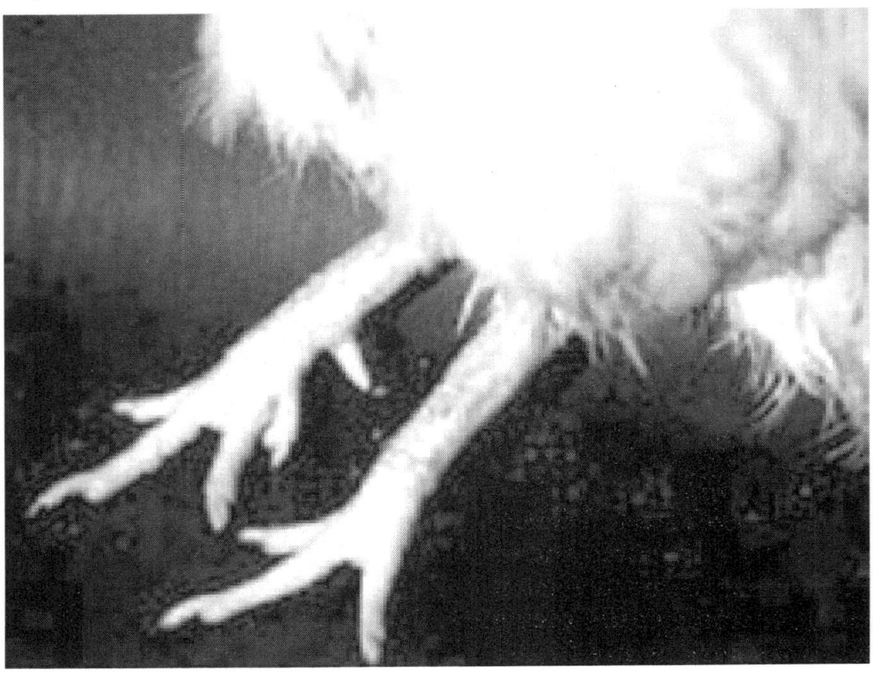

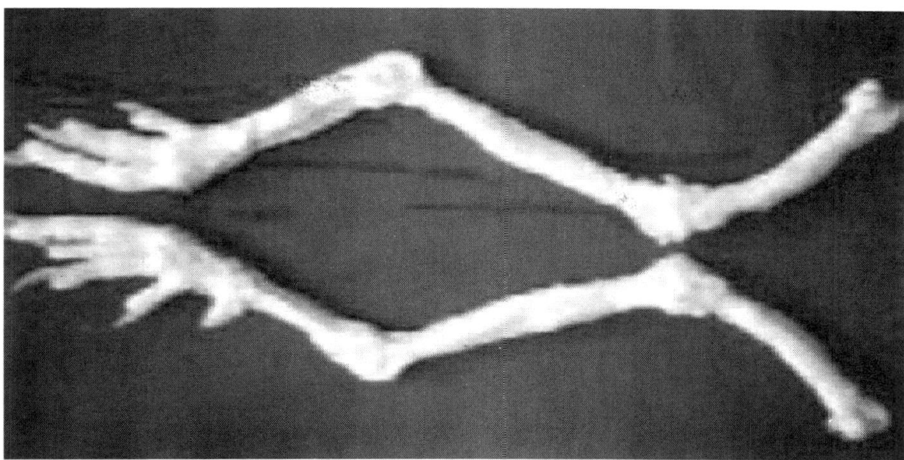

341.342.Osteopetrosis is a neoplastic disease, aetiologically related to the L/S group of viruses. It is characterized by a significant thickening of bone periosteum. The diaphyses of the tibia and/or tarsometatarsal bones are most commonly affected. Often, osteopetrosis is seen simultaneously with LL in the same bird.

Adenocarcinomatosis

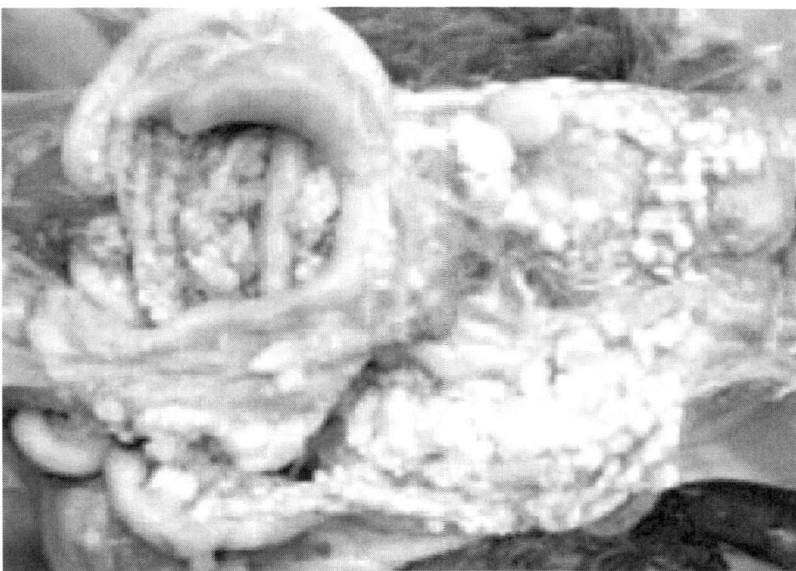

343.The intestinal and reproductive tracts are primarily affected. Quite often, the neoplasms invade the peritoneum and other serous coats. Macroscopically they appear as numerous disseminated thick nodes of a various size (with a diameter from several mm to 1 cm).

344.Sometimes, the tumours are cystic formations (cystadeno-carcinoma).

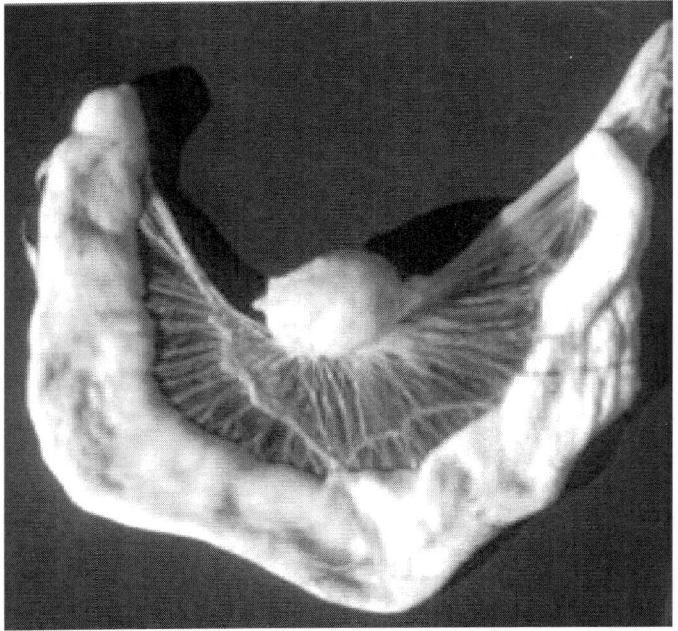

345.The leiomyoma of the mesosalpinx is a common tumour in hens. Usually, it is located in the ventral ligament of the oviduct. Its size varies from small to large (several cm in diameter), smooth, thick, sometimes highly vascularized nodes.

Parasitic Diseases

Coccidiosis

346. Coccidiosis is a common protozoan disease in domestic birds and other fowl, characterised by enteritis and bloody diarrhoea. The intestinal tract is affected, with the exception of the renal coccidiosis in geese. Clinically, bloody faeces, ruffled feathers, anaemia, reduced head size and somnolence are observed.

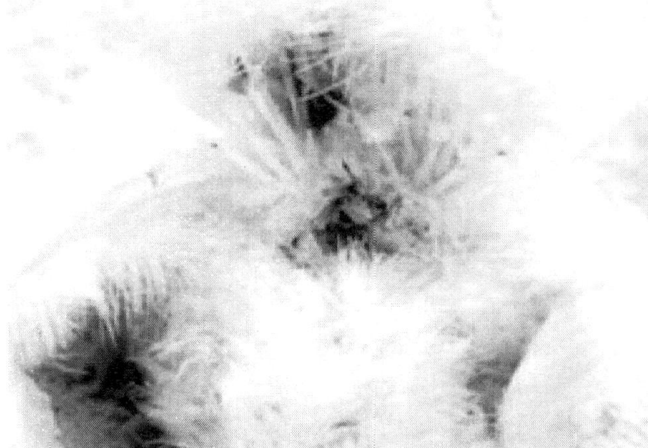

347. The area around the vent is stained with blood. The infection is realised by a faecal-oral route. After ingestion of sporulated (infective) oocysts, sporozoites are released that enter asexual and sexual cycles of development resulting in the emergence of thousands of new oocysts in the intestines. Oocysts are distributed by faeces. Soon, they sporulate and become infective for chickens.

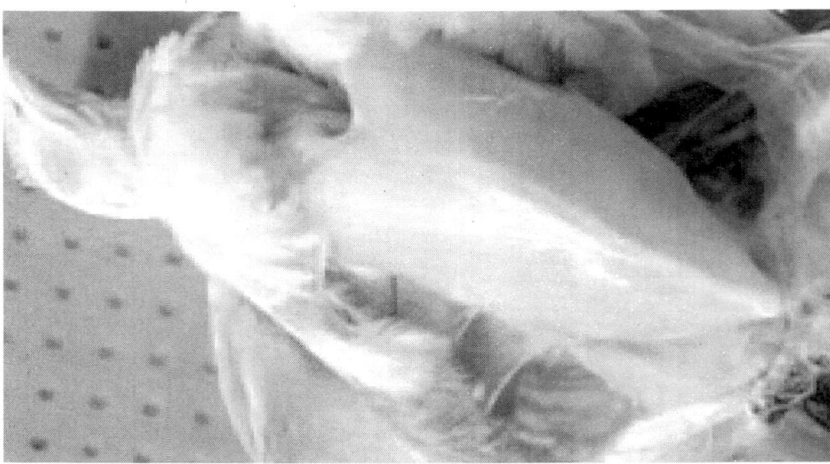

348. The intestinal lesions provoked by coccidia, are due to injury of the epithelial cells of the mucous coat where the parasites are developed and multiplied. The oocysts exist in the litter in premises and are distributed by clothes, shoes, dust, insects etc. Pathoanatomically, dehydration and a high degree of anaemia of the body and viscera are discovered.

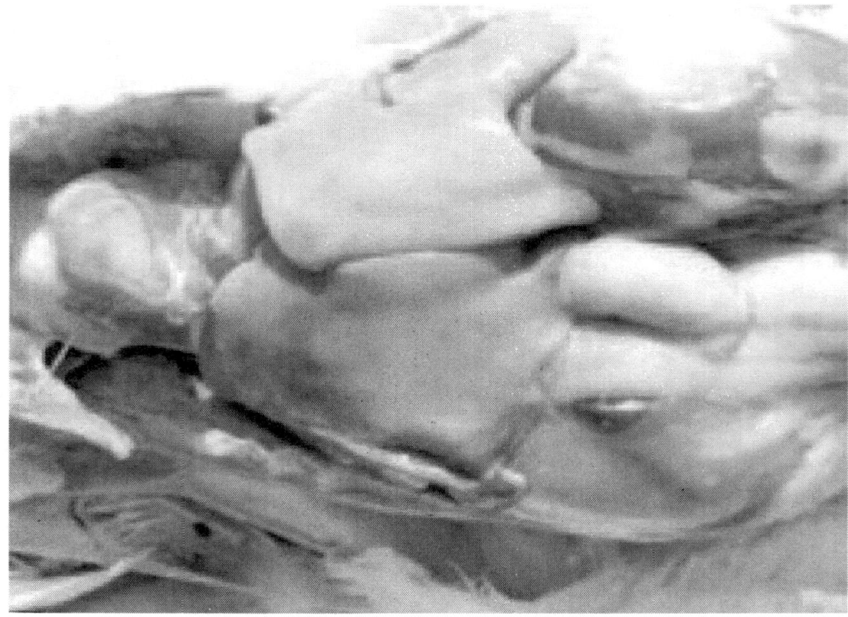

349. Anaemic appearance of internal organs. The wet litter and the heat in premises favour of the sporulation and therefore, the outbreak of coccidiosis.

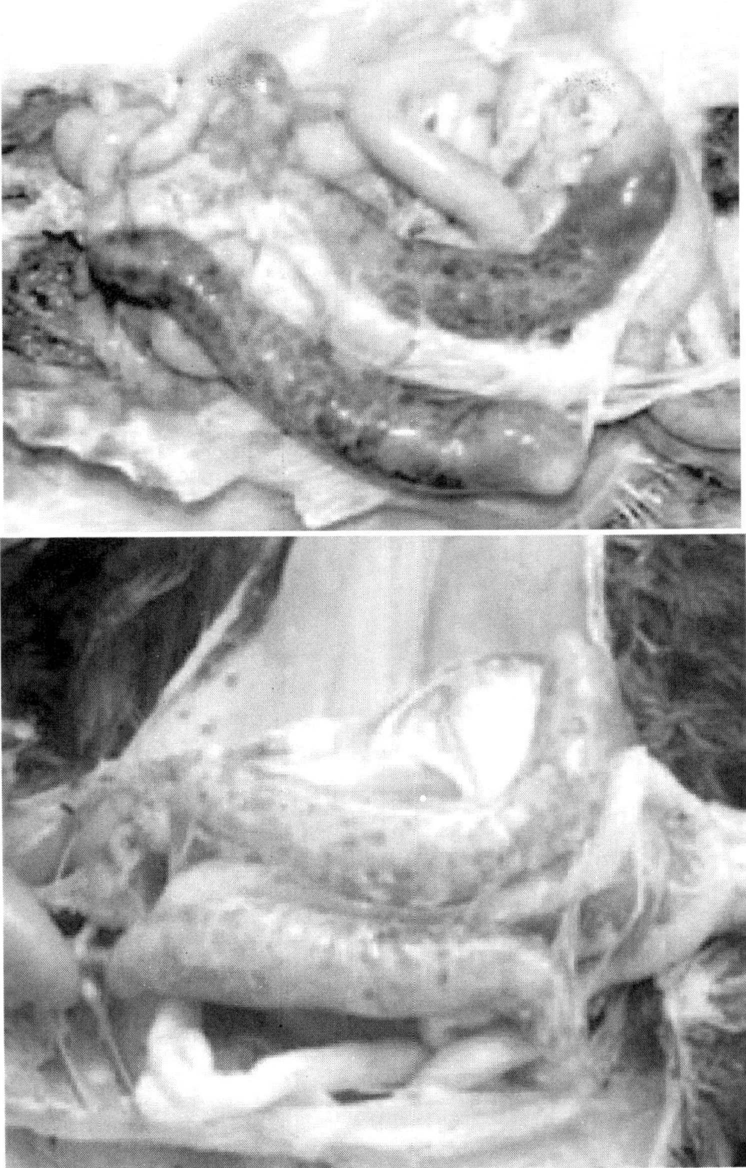

350.351. Depending on the localisation of lesions in intestines, the coccidioses are divided into caecal, induced by E. tenella, and small intestinal, induced by E. acervulina, E. brunetti, E. maxima, E. mitis, E. mivati, E. necatrix, E. praecox and E. nagani. In caecal coccidiosis, a marked typhlitis is present and haemorrhages are seen through the intestinal wall.

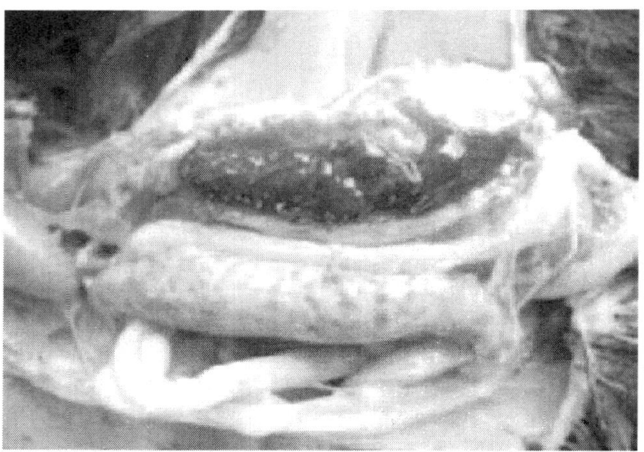

352. The caeca are filled with fresh or clotted blood.

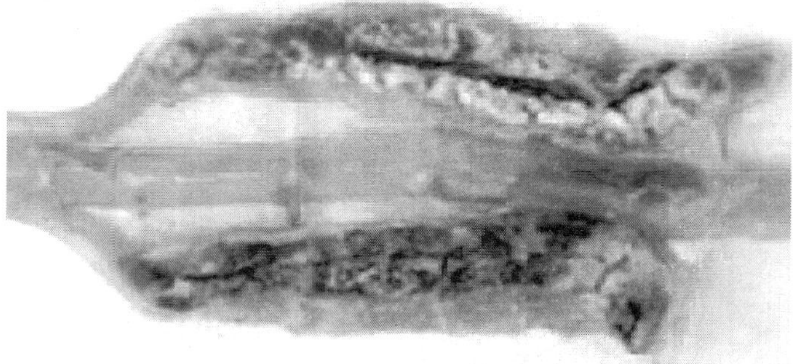

353. A later stage, the caecal content becomes thicker, mixed with fibrinous exudate and acquires a cheese like appearance.

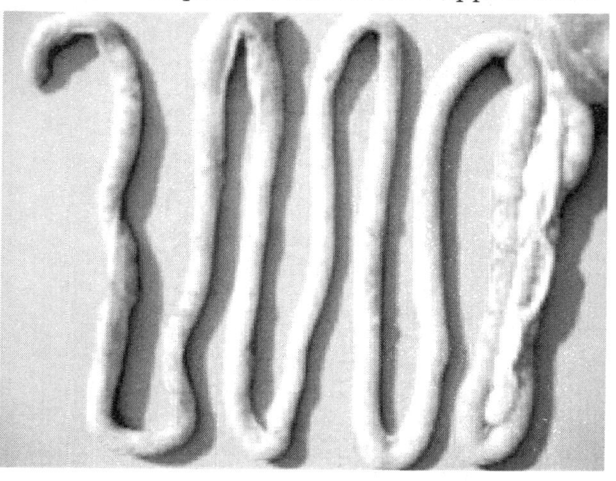

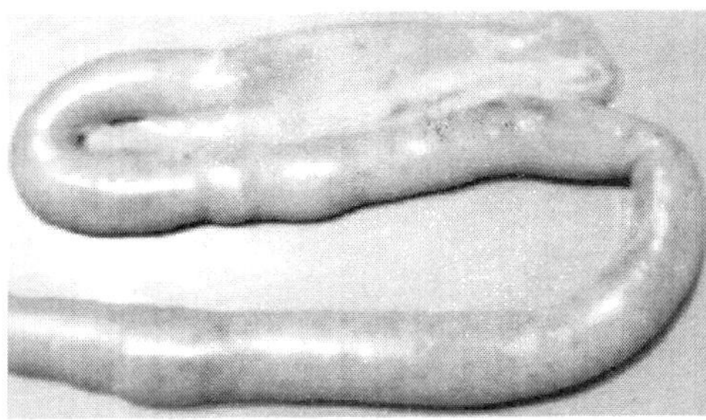

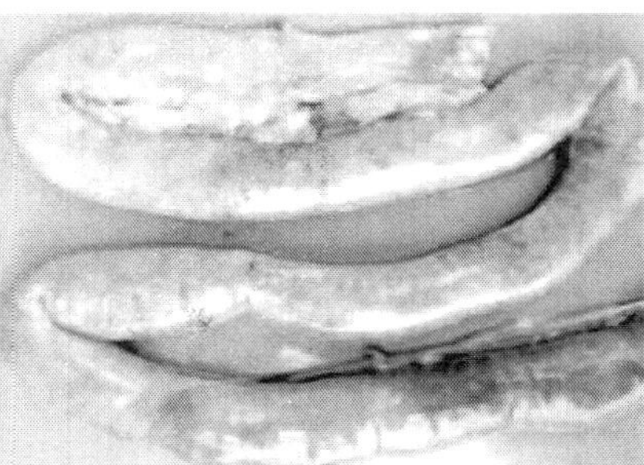

354.355.356. In small intestinal coccidioses, depending on the eimeria species, haemorrhages with various intensities in different parts along the intestine are observed. In many instances, the haemorrhages are petchial and could be seen through the intestinal wall.

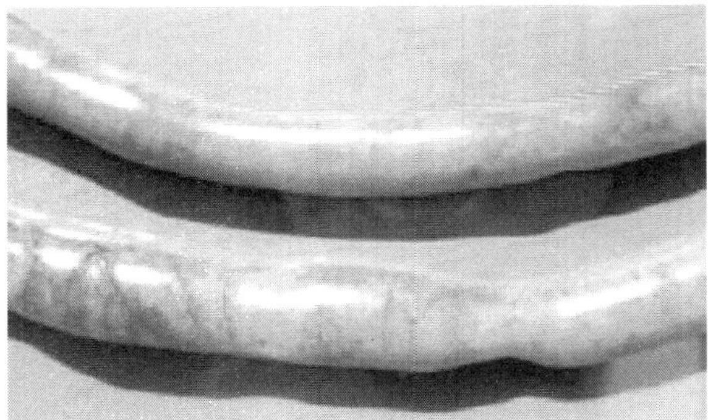

357. Sometimes, a reation of the intestinal lymphoid tissue is present.

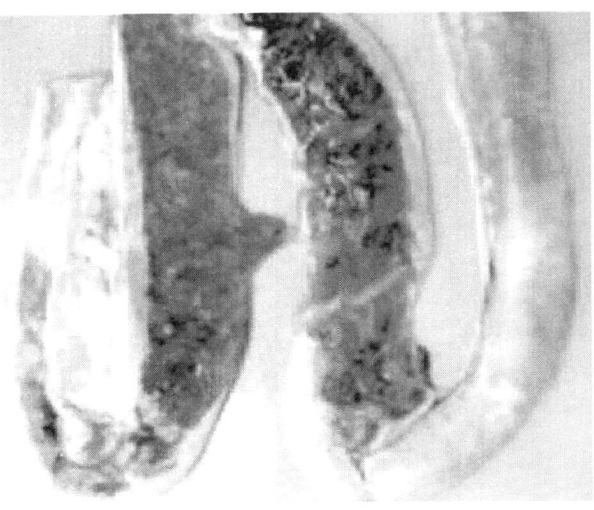

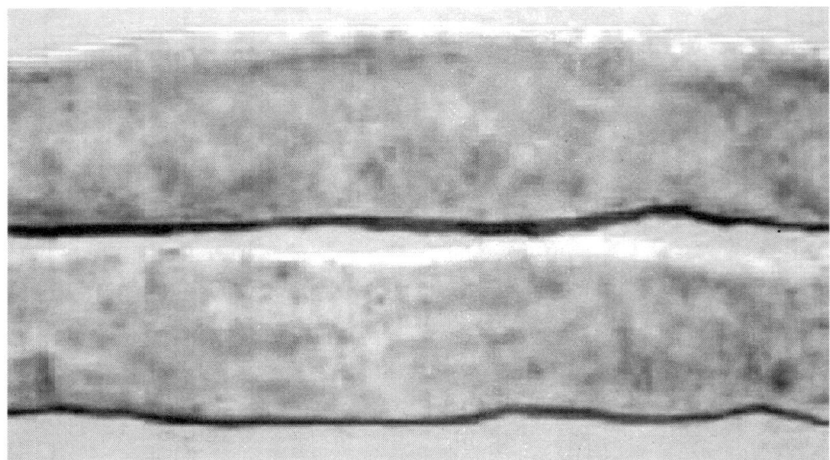

358.359. The content is mixed with fresh or clotted blood, and the mucous coat is mottled with multiple petechial or larger haemorrhages.

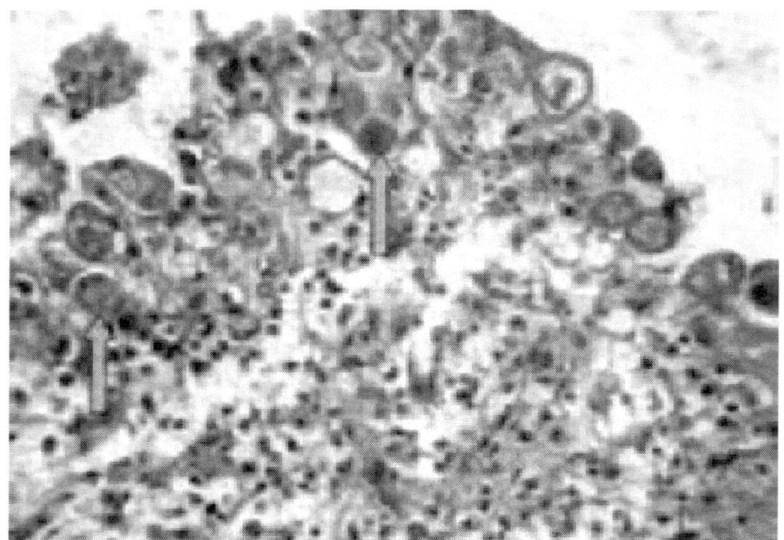

360. Histologically, Eimeria organisms at a various stage of development are detected in the epithelial intestinal cells. The diagnosis is made upon the results of the complex evaluation of the clinical picture, the macroscopic lesions, imprint preparations, histological study and flotation. Coccidioses should be differentiated from NE, UE and histomonosis (typhlohepatitis). Treatment - sulfonamides are widely used: sulfadimethoxine, sulfaquinoxaline, I sulfamethazine, but they should not I be used in layer hens. The supplementation I of vitamins A and K promotes the recovery.

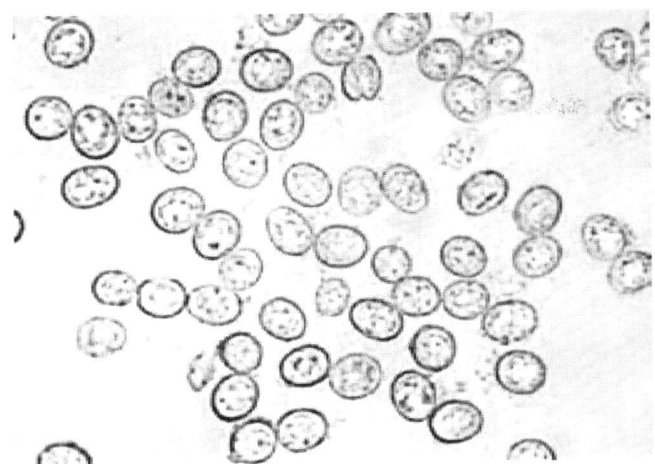

361. The microscopic examination of a native preparation of intestinal content or superficial mucosal layer reveals a significant number of oocysts in one observation field. Prevention. The use of coccidiostatics with forages on a rotation basis is the most extensively used means. The immunisation against coccidiosis with commercial vaccines is used in broiler breeder flocks. If the chickens are exposed to the natural effect of a moderate number of oocysts in their environment, they develop immunity to the respective parasitic species.

Histomonosis

362.. The histomonosis is a protozoan disease, caused by Histomonas meleagridis, and characterised by necrotising lesions affecting the liver and the caeca. Clinically, sulfur-yellow coloured faeces and depression are observed. A characteristic feature is the blackening of the skin of the head (blackhead), due to cyanosis.

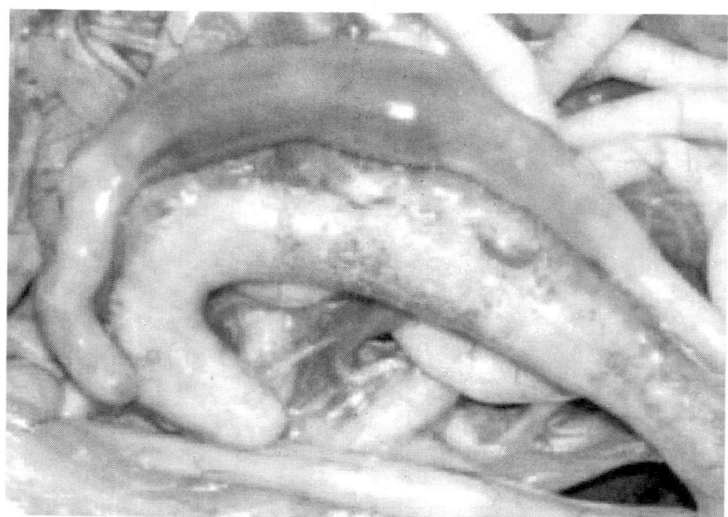

363. Pathoanatomically, bilateral enlargement of caeca with thickening of walls is observed. The aetiological agent is Histomonas meleagridis, a polymorphic flagellate that is present as flagellate in caeca and as amoeba in tissues. The trophozoites survive for several hours in the environment but in Heterakis eggs, they remain infective for more than a year.

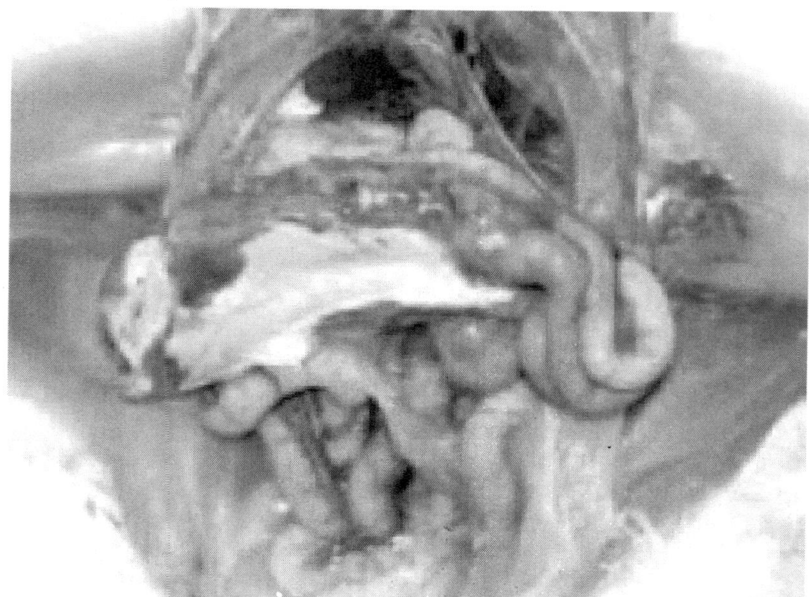

364. Often, the occurring typhlitis is the cause for adhesive peritonitis.

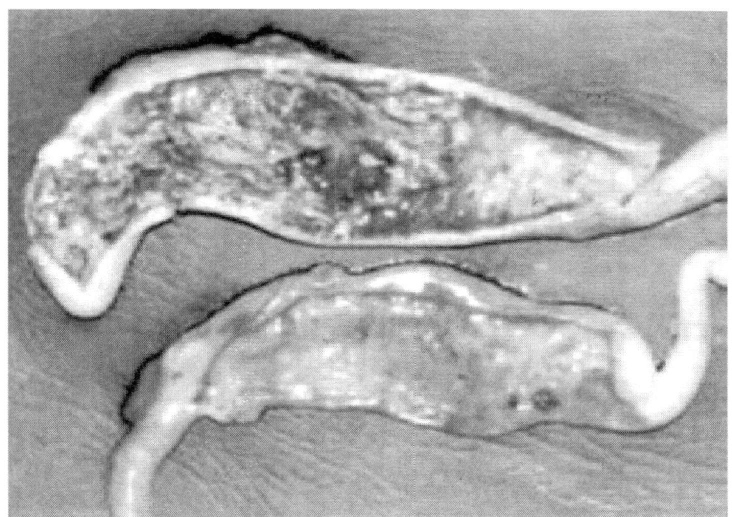

365. Susceptible species are turkeys, chickens, pheasants, rock partridges, guinea fowl, and geese. The turkeys are the most vulnerable between 3 and 12 weeks of age and chickens between 4 and 6 weeks of age. The caecal mucosa is usually ulcerated.

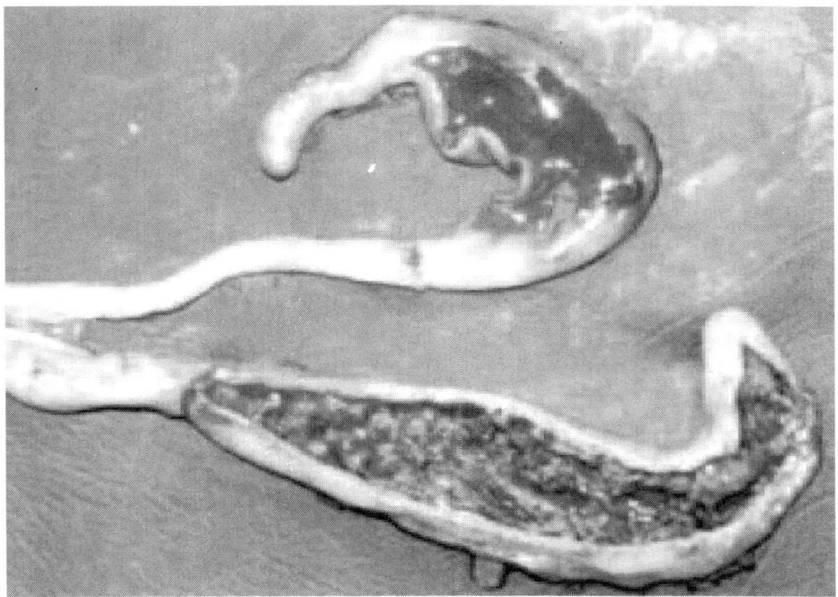

366.. The main vector is Heterakis gallinarum through the eggs, respectively the larvae, where Histomonas meleagridis forms are found. Some wild birds could also serve as vectors. The caecal content is often mixed with blood.

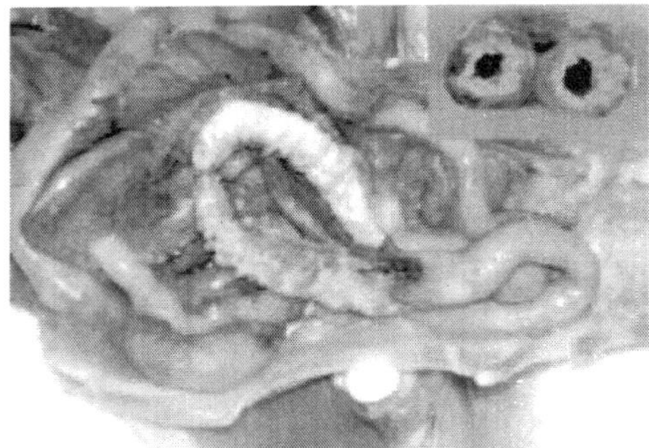

367. In older cases, crusts of dense caseous masses are formed into the carca that thicken this intestinal wall and reduce the lumen.

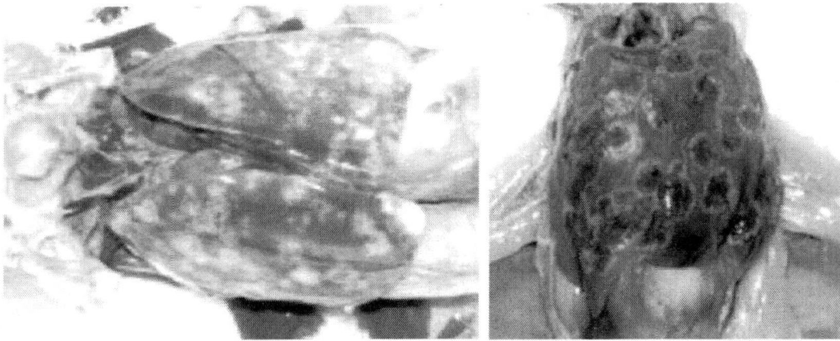

368.369. Earth worms are mechanical vectors of H. gallinarum larvae. The main reservoirs of infection are hens and chickens. The morbidity rate amounts to 90% and the mortality rate to 70%. In the liver, irregularly outlined coagulation necroses with various size and colour, are observed.

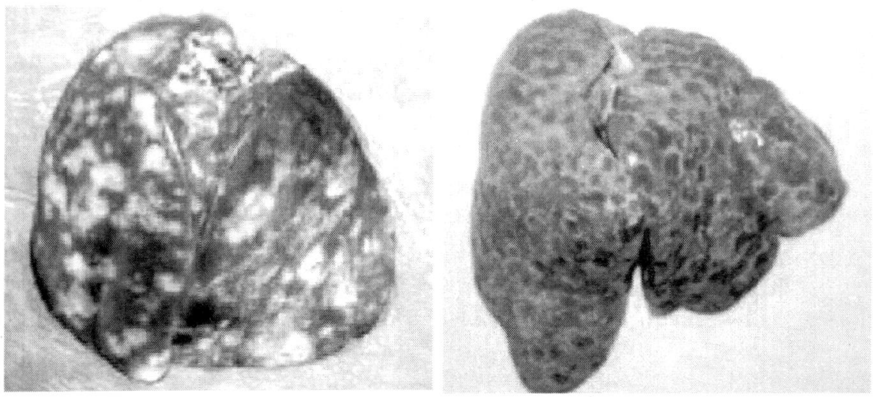

370.371. Usually, necroses represent yellowish to grey or red (haemorrhagically infarcted), well delineated oci with diameter of about

1-2 cm. Diagnosis - it is made on the basis of the typical macroscopic lesions. When necessary, a histological study and phase-contrast microscopy of native preparations could be performed.

Histomonosis should be differentiated from UE, coccidiosis and alimentary tract trichomonosis (Trichomonas gallinarum), where not counting the caeca, lesions are also present in the last third of small intestine.

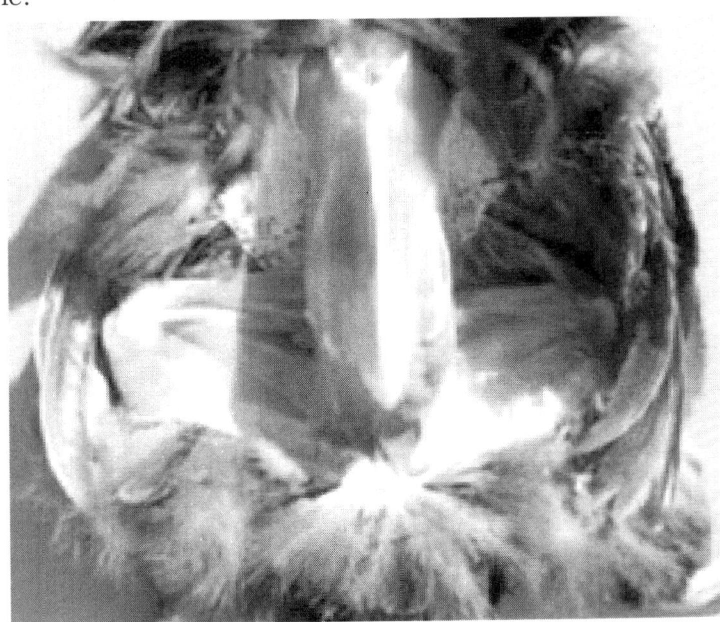

372.Ascaridiosis is one of the most prevalent helminthoses in fowl. It is caused by various species from the Ascaridia genus. The ascarids have a direct life cycle. Sometimes, it could involve paratenic hosts (earth worms). Infected birds are progressively emaciated, anaemic and sometimes diarrhoeic.

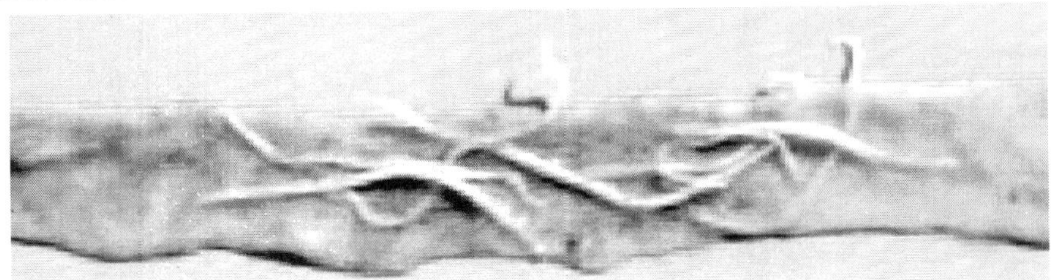

373.Pathoanatomically, haemorrhages of various intensities are found out in intestinal mucosa, catarrhal haemorrhagic enteritis and the parasites themselves are also observed.

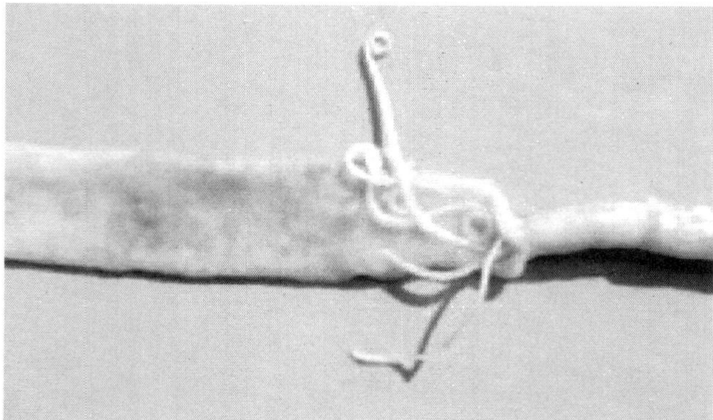

374.In cases of extensive invasion, the ascarids block the intestinal lumen and could cause a complete obstruction. The treatment and the control are realized through regular pathoana-tomical and coproovoscopic studies and performance of therapeutic and protective dehelminthizations.

375.Raillietinosis is a cestodosis characterized by diarrhoea (sometimes bloody) during the acute stage and emaciation to cachexia and anaemic during the chronic stage. It is caused by some representatives of the Raillietina genus that parasitize in various areas of the small intestine. The usual inter¬mediate hosts are ants or other insects. Pathoanatomically, haemorrhages with various intensities in the intestinal mucosa, catarrhal haemorrhagic enteritis and the parasites themselves are found out throughout the gross examination. The treatment and the control are done by dehelminthization of all birds in the affected farm.

Bibliography

Abbas, M.A.: *Effect of Organic Amendments and EM on Crop Production in Pakistan,* SP, Brazil. Pub. USDA. Washington, D.C., 19991.

Ahn., D.U. : *Effects of Post-Mortem Time Before Chilling and Chilling Temperatures on Water Holding Capacity and Texture of Turkey Breast Muscle,* Poultry Science, 1997.

Ball, W.: *Egg Quality Guidelines for the Australian Egg Industry,* AECL Publication, UK, 2004.

Bhosale, Dinesh T. *: Handbook of Poultry Nutrition,* International Book, 2004.

Cox, N. A. : *Relationship Between Aerobic Bacteria, Salmonella, And Campylobacter on Broiler Carcasses,* Journal of Food Protection, 1997.

Cramer, C.: *Sustainable Farming Connection: Where Farmers Find and Share Information,* Sustainable Farm Publishing, US, 1997.

Dennis Wages: *Biosecurity in the Poultry Industry,* International Book, Delhi, 2006.

Elson, H.A.: *Poultry Production Systems, Behaviour, Management and Welfare,* CAB International, NY, 1992.

Ferguson, M.W.J.: *Egg Incubation, Its Effects on Embryonic Development in Birds and Reptiles,* Cambridge University Press, UK, 1991.

Foreman, P.: *The Chicken Tractor: The Permaculture Guide to Happy Hens and Healthy Soil-All New Straw Bale Edition,* Good Earth Publications, USA, 2002.

Garnsworthy, P.C.: *Recent Developments in Poultry Nutrition,* University Press, India, 1999.

Gillespie, J.R.: *Modern Livestock and Poultry Production 7th edition,* Thomson Delmar Learning, 2004.

Hernandez, J.: *Optimum Egg Quality: A Practical Approach,* 5M Publishing, U.K, 2007.

Keith Wilson N.D.P: *A Handbook of Poultry Practice,* Agrobios, Delhi, 2000.

Leeson, S & Summers, J.D.: *Scott's Nutrition of the Chicken,* University Books, UK, 2001.

Leeson, S. and J.D. Summers: *Commercial Poultry Nutrition*, Nottingham University Press, Delhi, 2008.

Mandal, A.B. : *Nutrition and Disease Management of Poultry*, International Book Distributing Co, Delhi, 2004.

Mead, G.: *Food Safety Control in the Poultry Industry,* Woodhead Publishing Limited, Abington Hall, Abington, Cambridge, 2006.

Nicholls, C.: *The Workboot Series: The Story of Eggs in Australia,* Kondinin Group, Cloverdale W.A., 2005.

North, M.O.: *Commercial Chicken Meat and Egg Production,* Kluwer, USA, 2001.

Owen, W Powell : *Poultry Farming and Keeping*, Biotech Books, Delhi, 2005.

Panda, A K ; S V Rama Rao and M R Reddy: *Growth Promoters in Poultry : Novel Concepts*, International Book Dist, Delhi, 2008.

Pathak, N.N. : *Nutrition and Disease Management of Poultry*, International Book Distributing Co, Delhi, 2004.

Randall, C.J.: *Color Atlas of Diseases and Disorders of the Domestic Fowl and Turkey,* Iowa State University Press, UK, 1991.

Reddy, M R : *Growth Promoters in Poultry : Novel Concepts*, International Book Dist, Delhi, 2008.

Sharma, R.P. : *Poultry Production in India*, Indian Council of Agricultural Research, Delhi, 2007.

Sharma, S.R. : *Poultry Production in India*, Indian Council of Agricultural Research, 2008.

Shukla, Rakesh Kumar : *Handbook of Poultry Diseases : A Bedside-Guide*, International Book Dist, Delhi, 2006.

Singh, Ram Prakash : *Modern Livestock and Poultry Production*, Biotech Books, Delhi, 2008.

Sreenivasaiah, P V : *Scientific Poultry Production : A Unique Encyclopaedia*, International Book Dist, Delhi, 2006.

Tsushima, T. : *Survey of Bacterial Contimanation and Microbiological Control at a Poultry Slaughterhouse.* Journal of the Japan Veterinary Medical Association, 1996

Vegad, J L : *Poultry Diseases : A Guide for Farmers and Poultry Professionals*, International Book Dist, Delhi, 2008.

Verma, S. S. : *Microbiological Changes on Chicken Carcasses During Processing*, Indian Journal of Poultry Science, 1989.

Whitehead, C.C.: *Bone Biology and Skeletal Disorders in Poultry,* Carfax Publishing Company, U.K., 1992.

Young, M.: *Controlling Newcastle Disease in Village Chickens,* ACIAR, US, 2002.

Index

A

B

C

D

E

F

G

H

I

❑❑❑